The Massey Lectures Series

The Massey Lectures are co-sponsored by CBC Radio, House of Anansi Press, and Massey College in the University of Toronto. The series was created in honour of the Right Honourable Vincent Massey, former governor general of Canada, and was inaugurated in 1961 to provide a forum on radio where major contemporary thinkers could address important issues of our time.

This book comprises the 1989 Massey Lectures, "The Real World of Technology," broadcast in November 1989 as part of CBC Radio's *Ideas* series. The producer of the series was Max Allen; the executive producer was Bernie Lucht. The author included a new Preface, four new chapters, and a Coda for this 1999 edition of the book.

Ursula M. Franklin

In 1948 Ursula M. Franklin received her PhD in experimental physics in Berlin. She came to Canada the following year and began a distinguished scientific career. She joined the University of Toronto's Department of Metallurgy and Materials Science, becoming a full professor in 1973. Dr. Franklin is a Companion of the Order of Canada and a Fellow of the Royal Society of Canada. She has received honorary degrees from many Canadian universities. In 1984 she became the first woman to be honoured with the title of University Professor by the University of Toronto. In 1987 Dr. Franklin was awarded the Elsie Gregory McGill memorial award for her contributions to education, science, and technology. In 1989 she received the Wiegand Award, and in 1990 was awarded the Order of Ontario.

As a Quaker, she has been actively involved in work for peace and justice, international understanding, and women's issues. In 1995 the Toronto Board of Education named a new public school in her honour. In recognition for her humanitarian efforts, Franklin was awarded the Pearson Peace Medal in 2001.

THE REAL WORLD OF
TECHNOLOGY

Ursula M. Franklin

{ REVISED EDITION }

ANANSI

First published in 1990 by CBC Enterprises/les Entreprises Radio-Canada
Published in 1992 by House of Anansi Press Ltd.
Revised Anansi edition published in 1999

This edition published in 2004 by
House of Anansi Press Inc.
110 Spadina Avenue, Suite 801
Toronto, ON, M5V 2K4
Tel. 416-363-4343
Fax 416-363-1017
www.anansi.ca

Distributed in Canada by
HarperCollins Canada Ltd.
1995 Markham Road
Scarborough, ON, M1B 5M8
Toll free tel. 1-800-387-0117

Distributed in the United States by
Publishers Group West
1700 Fourth Street
Berkeley, CA 94710
Toll free tel. 1-800-788-3123

CBC and Massey College logos used with permission

House of Anansi Press is committed to protecting our natural environment. As part
of our efforts, this book is printed on Rolland Enviro paper: it contains 100%
post-consumer recycled fibres, is acid-free, and is processed chlorine-free.

10 09 08 07 06   5 6 7 8 9

LIBRARY AND ARCHIVES CANADA CATALOGUING IN PUBLICATION DATA

Franklin, Ursula M., 1921–
The real world of technology
(CBC Massey lectures series)
Rev. ed.
ISBN 0-88784-636-X
1. Technology — Social aspects. I. Title II. Series.
T14.5.F73 1999   303.48′3   C98-933052-4

Library of Congress Control Number: 2006920453

Cover design: Bill Douglas at The Bang
Text design: Tannice Goddard

Canada Council   Conseil des Arts
for the Arts   du Canada

ONTARIO ARTS COUNCIL
CONSEIL DES ARTS DE L'ONTARIO

*We acknowledge for their financial support of our publishing program
the Canada Council for the Arts, the Ontario Arts Council, and the Government of Canada
through the Book Publishing Industry Development Program (BPIDP).*

Printed and bound in Canada

Contents

PREFACE TO THE
NEW EDITION

It has been a decade since *The Real World of Technology* was originally prepared for the CBC Massey Lectures, a decade during which technological changes have driven major global political and economic changes.

Revisiting *The Real World of Technology* in order to illustrate the new reality of a technological world as well as its trends and impacts, I feel it is best to let the 1989 lectures stand unamended and to add four new chapters on some of the more recent facets of the emerging picture to this edition. Much to what was said a decade ago about the real world of technology remains valid since much of what appears new can be seen as an extension of earlier developments. Many dominant trends and problems discussed in 1999 can be explained as the consequences of previously delineated configurations and dynamics. I will continue to define "technology" as "practice" — as the way things are done around here — and will emphasize how the practices and their contexts have changed.

The first new chapter (chapter seven) will deal with communications technologies, both ancient and modern. From the invention of writing to the use of the Internet, the way in which knowledge is kept, transmitted, or shared has structured the perception of what is real, as well as what is possible or desirable.

The next new chapter explores and illuminates some novel aspects of the new electronic technologies as they reshape our use and experience of time. Chapter nine will then offer a model of the technologically changed configurations of space and their political consequences.

Finally, chapter ten will focus on the impact of these new technological practices on human ties, on work and community, on governance, citizenship, and the notion of individual and collective responsibility.

Throughout these new chapters I want to stress how new technologies, new ways of doing things, have pushed against the physical and social boundaries of space and time. These activities have altered profoundly the relationships of people to nature, to each other, and their communities.

It is with profound gratitude that I acknowledge again the support of my family and my communities, particularly the hospitality of Massey College and the friendship of its Master, John Fraser.

I would like to dedicate this edition to my three young grandsons, who may never know how much I worry that the world in which they will live may be one of justice, peace, and beauty.

PREFACE TO THE 1990 EDITION

The CBC's annual Massey Lectures are a time-honoured Canadian tradition, and each Massey lecturer must have had days, as I did, when this tradition felt like a weighty burden. The opportunity to deliver the 1989 Massey Lectures as six *public* talks lightened this burden for me and added greatly to the pleasures and the rewards of this assignment. The extensive discussions after each lecture, the correspondence resulting from these exchanges, and the letters I received after the radio broadcasts showed me just how broadly-based and profound are the concerns in this country about the social and moral impacts of technology. The comments reflected a considerable diversity of approaches, a readiness for new perspectives, and a remarkable willingness to listen and to engage in discourse.

These responses strengthened one of my images of a peaceful world: a society that might work somewhat like a pot-luck supper, where everyone contributes and everyone receives, and where a diversity of offerings is essential. (Just imagine a pot-luck to which everyone brought potato salad!). In such a world there would be no one who could not contribute their work and care — and no one who could not count on receiving nourishment and fellowship. I hold this vision with increased confidence.

It would not have been possible for me to carry out the work reflected in these lectures without the support, friendship, and stimulation of many people. I draw much strength from the support of my family and from the tenets of Quakerism. For more than a quarter of a century the women's peace movement has been so much a part of my life that I often do not know whether they or I speak through my words. I would not have dared to undertake the preparation of the Massey Lectures without knowing that I could rely on the help and support of the CBC *Ideas* team; to my friend and colleague Max Allen I owe a special debt of gratitude for his part in shaping these lectures in their spoken and written forms. The co-sponsorship of the lectures by Massey College at the University of Toronto assisted greatly in the public delivery of the series and in the ensuing discussions. To the college and its former master, Ann Saddlemyer, go my sincere thanks.

THE REAL WORLD OF
TECHNOLOGY

I

I start from the premise that we are living in a very difficult, very interesting time, a time in which a major historical period is coming to a convoluted end. I think we live in a time in which the social and political upheaval is as great or greater than it was at the time of the Reformation. And so I would like to do a bit of orienteering and map-making so the discourse in which we all have to engage can be conducted in a common language.

As I see it, technology has built the house in which we all live. The house is continually being extended and remodelled. More and more of human life takes place within its walls, so that today there is hardly any human activity that does not occur within this house. All are affected by the design of the house, by the division of its space, by the location of its doors and walls. Compared to people in earlier times, we rarely have a chance to live outside this house. And the house is still changing; it is still being built as well as being demolished. In these lectures, I would like to take you through the house, starting with the foundation and then examining with you the walls that have been put up or taken down, the storeys and turrets that have been added, the flow of people through the house — who can come in, who can go into particular spaces.

In the past, I have often spoken about the social impact of technology in terms of apprehension and foreboding, but this is not my purpose here. *My interest is in contributing to clarity.* I want to know as much as possible about the house that technology has built, about its secret passages and about its trapdoors. And I would also like to look at technology in the way C. B. Macpherson looked at democracy[1] — in terms of the real world. Technology, like democracy, includes ideas and practices; it includes myths and various models of reality. And like democracy, technology changes the social and individual relationships between us. It has forced us to examine and redefine our notions of power and of accountability.

In this lecture, I would like to talk about *technology as practice*, about the organization of work and of people, and I would like to look at some models that underlie our thinking and discussions about technology. Before going any further, I should like to say what, in my approach, technology is not. Technology is not the sum of the artifacts, of the wheels and gears, of the rails and electronic transmitters. Technology is a *system*. It entails far more than its individual material components. Technology involves organization, procedures, symbols, new words, equations, and, most of all, a mindset.

In subsequent lectures, I will focus on technology as it has changed our realities of time and space. I will talk about planning and forecasting, and about the many attempts to predict the impact of technology. It will also be necessary to deal with the confusion that sometimes exists between the notion of the so-called objectivity of technology and the fact that, in a narrow sense, some of the outcomes of technological processes are predictable. It will be important to examine the human and the social impacts of technology, and we must talk about the changing nature of experience, as well as about the fragmentation of knowledge and work. We need to examine the new social class of experts, as well as the changing nature of community and constituency that has been brought about by technological systems. Technology also needs to be examined as an agent of power and control, and I will try to show how much modern technology drew from the prepared soil of the structures of traditional institutions, such as the church and the military. I will also talk about scale and complexity and about the modem technologies of management and government. I would also like to talk about some of the technologies that are of particular interest to me, such as communications technologies and computers. I will trace in one of the lectures the life-cycle and pattern of development of specific technologies from innovation and incipient development, through to the point when the technologies become entrenched in the social landscape. And I would like to touch on the place of the natural environment when examining technologies as agents of change. I will trace back some of the themes, so that we might see what developments and social transformations will be needed for the real world of technology to become a healthy and sane habitat for human beings.

There exists a vast literature on all aspects of technology.[2] Recently, people have become particularly interested in technology's social impact. I myself am overawed by the way in which technology has acted to reorder and restructure social relations, not only affecting the relations between social groups, but also the relations between nations and individuals, and between all of us and our environment. To a new generation, many of these changed relationships appear so normal, so inevitable, that they are taken as given and are not questioned. Yet one can establish clear historical trends. In order to understand the real world of technology and cope with it, we need to have some knowledge of the past, as well as to give some thought to the future.

Altogether there seems to be a very drastic change in what it means today to be human — what it means to be a woman, a child, a man; to be rich or poor; to be an insider or an outsider — compared with what all this meant in the past. My father was born just before the turn of the century, and many of his approaches to life could be understood from knowing that he was a German intellectual, a member of an old family of a particular social class, coming from a particular region of the country. On the other hand, you would know very little about my son's approach to life and his values and attitudes by knowing that he is a photographer, born in the late 1950s in Toronto. Technology has muddled or even destroyed the traditional social compass.

It is my conviction that nothing short of a global reformation of major social forces and of the social contract can end this historical period of profound and violent transformations, and give a manner of security to the world and to its citizens. Such a development will require the redefinition of rights and responsibilities, and the setting of limits to power and control. There have to be completely different criteria for what is permissible and what is not. Central to any new order that can shape and direct technology and human destiny will be a renewed emphasis on the concept of justice. The viability of technology, like democracy, depends in the end on the practice of justice and on the enforcement of limits to power.

Let me begin now, like any good academic, with definitions. To define technology in a global sense is really quite fraught with difficulties; the best minds among philosophers, historians, social scientists, and engineers have attempted to do it.[3] I will not compete with them. Rather, I would like to define, as we work through this, technology in its various aspects within the context in which they occur and within the context in which I discuss them. Like democracy, technology is a multifaceted entity. It includes activities as well as a body of knowledge, structures as well as the act of structuring. Our language itself is poorly suited to describe the complexity of technological interactions. The interconnectedness of many of those processes, the fact that they are so complexly interrelated, defies our normal push-me-pull-you, cause-and-consequence metaphors. How does one speak about something that is both fish and water, means as well as end? That's why I think it is better to examine limited settings where one puts technology in context, because context is what matters most.

As a beginning let's look at technology as practice. Kenneth Boulding, the author of *The Image* and many other influential books in the social sciences,[4] suggested that one might think of technology as *ways of doing something*. He pointed out that there is a technology of prayer as well as of ploughing; there are technologies to control fear as well as to control flood.

Looking at technology as practice, indeed as formalized practice, has some quite interesting consequences. One is that it links technology directly to culture, because culture, after all, is a set of socially accepted practices and values. Well laid down and agreed upon practices also define the practitioners as a group of people who have something in common because of the way they are doing things. Out of this notion of unifying practice springs the historical definition of "us" and "them." I think it is important to realize that the experience of common practice is one of the ways in which people define themselves as groups and set themselves apart from others. "Around here, that's how we do things," a group will say, and this is their way of self-identification, because "others" may do the same thing differently. A different way of doing something, a different tool for the same task, separates the outsider from the insider.

I once was invited to examine Chinese bronzes at the Freer Gallery of Art. The purpose of the gathering was to develop suggestions for research into the technology of Chinese bronze casting, to share knowledge and to avoid duplication. We were about six or eight, and I will never forget the scene. Most participants were art historians or museum people; I was the only researcher coming out of engineering. We were all looking at bronze fragments and we all had magnifying glasses, but my magnifying glass was different from the magnifying glasses that everyone else had. "They" had magnifying glasses that they put to their eyes and then lifted the object into proper viewing distance. I had a magnifying glass that I put on the object, and I manoeuvred my head into a good viewing position. They took one look at my magnifying glass and I was out. I was classified as an outsider. I was treated politely. A lot of good relationships came out of that meeting. Still, I can recall to this day that feeling of distance, the surprised looks. I knew I was respected, but I just wasn't one of them.

On another occasion I sat in the back of a large meeting room, listening to a long and boring discussion. I began to knit. A young woman came over, sat down next to me, and whispered, "I'd like to talk to you. You knit just like my mother." Of course, her mother was also German, and there is a German way of knitting.

The historical process of defining a group by their agreed practice and by their tools is a powerful one. It not only reinforces geographic or ethnic distributions, it also affects the gendering of work. When certain technologies and tools are predominantly used by men, then maleness becomes part of the definition of those technologies. It is for these deep-rooted reasons that it is so very difficult for women to enter what are now called "non-traditional" jobs. If engineers are male and maleness is part of engineering, then it's tough for men to accept women into the profession. The apparent ease with which women acquire the knowledge necessary to practise only seems to increase the perceived threat to the male practitioners. And so year after year, engineering faculties go through initiation procedures that are crude, sexist, and obscene in order to establish that the profession is male, even if some of the practitioners are not.

The common practice that a particular technology represents, in addition to leading to an identification with culture and gender, can also lead to the "right" of the practitioners to an exclusive practice of the technology. This is how the professions were born; clergy, doctors, lawyers, engineers, and social workers all claimed the exclusive right to certain tools and to certain technologies.

Another facet of the concept of technology as practice is the fact that the practice can define the content. I spoke earlier of Kenneth Boulding's remark that there is a technology of prayer as well as a technology of ploughing. The sacred books of most religions lay out the practices of prayer quite precisely, and that laying down of the practice means that other forms of worshipful activities, however deeply they may be felt, cannot be considered legitimate prayer. For instance, the playing of or listening to a particular piece of music may very well be felt as a deep plea for deliverance, but it isn't prayer. One has to keep in mind how much the technology of doing something defines the activity itself, and, by doing so, precludes the emergence of other ways of doing "it," whatever "it" might be. This has been so historically but it is even more so today, because so many activities are technologically structured.

It becomes so easy and seemingly objective to define the content by the way something is being done or prescribed to be done. Teaching, for instance, is now a clearly circumscribed activity that takes place in a particular location and is conducted by particularly trained or ordained practitioners, and whatever somebody may teach you in working together with you, it isn't the kind of learning for which you ever get a credit. I think it's important to realize that technology defined as *practice* shows us the deep cultural link of technology, and it saves us from thinking that technology is the icing on the cake. Technology is part of the cake itself.

Let's distinguish the two ways in which technology has developed. In the first place there are work-related technologies. Work-related technologies make the actual practice easier. Take, for instance, the substitution of electric typewriters for mechanical ones; this is indeed a work-related technological improvement. Secondly, there are control-related technologies, those developments that do not primarily address the process of work with the aim of making it easier, but try to increase control over the operation. Think of a word processor. A freestanding word processor is indeed work-related technology. But link those word processors into a work station — that is, into a system — and the technology becomes control-related. Now workers can be timed, assignments can be broken up, and the interaction between the operators can be monitored. Most modern technological changes involve control and thus new control-related applications have increased much faster than work-related ones.[5]

It is not difficult to understand the difference between control- and work-related technologies, but I would now like to introduce a concept that may be somewhat more difficult to grasp. I want to distinguish between two very different forms of technological development. The distinction we need to make is between *holistic technologies* and *prescriptive technologies*.[6] Again, we are considering technology as practice, but now we are looking at what is actually happening on the level of work. The categories of holistic and prescriptive technologies involve distinctly different specializations and divisions of labour, and consequently they have very different social and political implications. Let me emphasize that we are not asking *what* is being done, but *how* it is being done.

Holistic technologies are normally associated with the notion of craft. Artisans, be they potters, weavers, metal-smiths, or cooks, control the process of their own work from beginning to finish. Their hands and minds make situational decisions as the work proceeds, be it on the thickness of the pot, or the shape of the knife edge, or the doneness of the roast. These are decisions that only they can make while they are working. And they draw on their own experience, each time applying it to a unique situation. The products of their work are one of a kind. However similar pots may look to the casual observer, each piece is made as if it were unique. Here are a few lines from the poem "The Land" by Vita Sackville-West, which convey the meaning of holistic technologies: "All craftsmen share a knowledge. They have held reality down, flattened to a bench; cut wood to their own purpose, compelled the growth of pattern with the patient shuttle. Control is theirs."[7] Using holistic technologies does not mean that people do not work together, but the way in which they work together leaves the individual worker in control of a particular process of creating or doing something.

A quote from Melville Herskovits, taken from his *Economic Anthropology*,[8] is also helpful. He points to the often very sophisticated specialization one finds historically in various societies, and he writes ". . . certain men and woman [sic] specialize, not only in one technique, but in a certain type of product, as, for instance, where one woman will devote her time to the production of pots for everyday use and another will make pottery exclusively for religious rites. It must again be stressed that, except under most unusual circumstances, we do not find the kind of organization where one woman characteristically specializes in gathering the clay, another in fashioning it, and a third in firing the pots; or, where one man devotes himself to getting wood, a second to roughly blocking out the proportions of a stool or figure, and a third to finishing it."

It is the first kind of specialization, by product, that I call holistic technology, and it is important because it leaves the doer in total control of the process. The opposite is specialization by process; this I call prescriptive technology. It is based on a quite different division of labour. Here, the making or doing of something is broken down into clearly identifiable steps. Each step is carried out by a separate worker, or group of workers, who need to be familiar only with the skills of performing that one step. This is what is normally meant by "division of labour."

This type of division of labour is most familiar to us as it arose in the Industrial Revolution in Britain. The factory system of the time resulted from large-scale applications of such divisions of labour.[9] However, this kind of division of labour is actually much older. We find it among the late Romans, whose Terra Sigillata pottery or Samian ware was produced by a prescriptively controlled technology. Items were essentially mass-produced to very close tolerances, and we have good written descriptions of the labour organization and of the technology, as well as examples of the artifacts themselves.[10] But even a thousand years earlier there was the production of Chinese bronze vessels, organized as a prescriptive technology *par excellence*, with clearly defined process-determined divisions of labour.[11]

The Chinese way of casting bronze — and it began well before 1200 BC — is indeed a *production* method. It is also unique to China, where later the same division of labour and the same method of production was used for the casting of iron. This resulted in the appearance of cast iron in China more than a thousand years before cast iron was produced in the West. Before that time, iron was mainly fashioned as wrought iron by a technology that is much more holistic. In fact, the making of wrought iron is almost the prototype of a holistic technology.

I'd like to take a moment to describe Chinese bronze-casting techniques, not only because I love Chinese bronzes and I have spent a lot of my professional life studying them, but also because Chinese bronze casting is such a magnificent example of prescriptive technologies and their social impact. Please don't think that considering the details of Chinese bronze casting has nothing to do with the topic at hand. Understanding the social and political impact of prescriptive technologies is, in my opinion, the key to understanding our own real world of technology.

Imagine, then, it is 1200 BC, the height of the Shang Dynasty. A large ritual vessel has to be cast — let's say a cauldron, a three-legged Ding, examples of which can be seen in the Royal Ontario Museum. First a full-size model of the Ding is made. It is usually made in clay, although it could be wood. Archaeologists have discovered lots of these models; such a model is a complete likeness of the vessel and all its decorations. From this model a mold is made. This is done by putting layers of clay — first very fine clay, then coarser material — onto the model and letting this coating dry. The mold is then carefully cut into segments and taken off the model in the way we take the peel off an orange. Because the mold is taken off in pieces one speaks about a "piece mold" process. The mold segments are then fired so they keep their shape and their decorations. They must be fired at temperatures that are higher than the temperature of molten copper or bronze, which the mold later contains. Consequently this casting technology became possible only in a civilization that had developed the techniques for producing high-fired ceramics.

Once the piece molds are fired, they are reassembled around a core, leaving a gap between the core and the mold large enough to receive the molten metal. The mold assembly has, of course, to include a means of pouring the liquid metal into that gap between core and mold as well as ways for the air that the liquid metal displaces to escape completely so that the casting is of good quality. Once the mold assembly is finished and properly positioned in a casting pit, the liquid bronze can be poured.

Up to this point in the process, essentially two main steps have been executed. The designer and model builder have constructed the model in a manner that allows the formation and the cutting away of the mold. This involves design expertise as well as a full knowledge of all subsequent steps in the process, because they all depend on the proper design of the model.

The next steps of building up the mold, of cutting it away, firing it, and reassembling it around the core in order to make it ready for casting, constitute a series of operations where the expertise is essentially that of pottery work.

The casting steps that follow the assembly of the mold require different expertise. Here the metal has to be prepared; the alloy has to be mixed in proper proportions and fused to a temperature high enough to allow a successful casting. Most, if not all, Chinese bronzes contain, in addition to tin, enough lead to make possible the casting of objects with very finely and elaborately designed surfaces. We are here talking about large castings. It is astonishing that towards the end of the Shang Dynasty, the Chinese cast cauldrons that weighed eight hundred kilograms or more. From technical studies, such as X-rays of the vessels, we know that they were essentially cast in one pour. This means that groups of metal workers were handling about a thousand kilograms of liquid bronze to cast a large vessel. These alloys melt above 1000°C. They were poured from crucibles; a large number of them had to be ready for pouring at approximately the same time.

Just imagine yourself in charge of such a labour force. And remember, these castings were not extraordinary events. The archaeological record shows that such castings were done routinely. The amount of material found, and the knowledge that this constitutes only a small fraction of what was produced, assures us of the presence of a large, coordinated production enterprise.

It was only when I considered in detail — as a metallurgist — what such a production enterprise would entail, that the extraordinary social meaning of prescriptive technologies dawned on me. I began to understand what they meant, not just in terms of casting bronze but in terms of discipline and planning, of organization and command.

Let's focus, for instance, on the need for precision, prescription, and control that such a production process develops. In contrast to what happens in holistic technologies, the potter who made molds in a Chinese bronze foundry had little latitude for judgement. He had to perform to narrow prescriptions. The work had to be right — or else. And what is right is laid down beforehand, by others.

When work is organized as a sequence of separately executable steps, the control over the work moves to the organizer, the boss or manager. The process itself has to be prescribed with sufficient precision to make each step fit into the preceding and the following steps. Only in that manner can the final product be satisfactory. The work is orchestrated like a piece of music — it needs the competence of the instrumentalists, but it also needs strict adherence to the score in order to let the final piece sound like music. Prescriptive technologies constitute a major social invention. In political terms, prescriptive technologies are *designs for compliance*.

When working within such designs, a workforce becomes acculturated into a milieu in which external control and internal compliance are seen as normal and necessary. Eventually there is only one way of doing something. The Chinese could probably not have imagined making bronze in any other manner, just as we can't imagine cars being manufactured in any other way than the way it's done today around the globe.

Bronze-making was not the only prescriptive technology in ancient China. Similar approaches are found in the making of warp-determined textiles and certain pottery productions. I've argued that the historically very early acculturation of Chinese people into prescriptive work processes must be regarded as a formative factor in the emergence of Chinese social and political thought and behaviour.[12] This includes the early emergence of a Chinese bureaucracy in its all-encompassing forms, the Imperial examinations, and the stress on *li* — the right way of doing something.

Today's real world of technology is characterized by the dominance of prescriptive technologies. Prescriptive technologies are not restricted to materials production. They are used in administrative and economic activities and in many aspects of governance, and on them rests the real world of technology in which we live. While we should not forget that these prescriptive technologies are often exceedingly effective and efficient, they come with an enormous social mortgage. The mortgage means that we live in a culture of compliance, that we are ever more conditioned to accept orthodoxy as normal, and to accept that there is only one way of doing "it."

As time went on more and more holistic technologies were supplanted by prescriptive technologies. After the Industrial Revolution, when machines began to be added to the workforce, prescriptive technologies spread like an oil slick. And today, the temptation to design more or less everything according to prescriptive and broken-up technologies is so strong that it is even applied to those tasks that should be conducted in a holistic way. Any tasks that require caring, whether for people or nature, any tasks that require immediate feedback and adjustment, are best done holistically. Such tasks cannot be planned, coordinated, and controlled the way prescriptive tasks must be.

When successful, prescriptive technologies do yield predictable results. They yield products in numbers and qualities that can be set beforehand, and so technology itself becomes an agent of ordering and structuring. (This aspect of technology is easily underestimated by those who see technology as mainly the application of scientific knowledge to real-life needs and problems.) The ordering that prescriptive technologies has caused has now moved from ordering *at* work and the ordering *of* work, to the prescriptive ordering of people in a wide variety of social situations.

For just a glimpse of the extent of such developments, think for a moment about the new "smart" buildings. Those who work in the buildings have a card with a barcode that allows them to get into the areas of the building where they have work to do but excludes them from anywhere else. Here we have what Langdon Winner, in his book *The Whale and the Reactor*,[13] calls so nicely "the digitalized footprints of social transactions," since the technology can be set up not only to include and exclude participants, but also to show exactly where any individual has spent his or her time. You could — just in a flight of fancy — imagine what would have happened if Adam and Eve had not lived in a garden but in a smart building. The divine designer would probably have arranged it so that they never saw apples. But, joking aside, prescriptive technologies eliminate the occasions for decision-making and judgement in general and especially for the making of *principled* decisions. Any goal of the technology is incorporated *a priori* in the design and is not negotiable.

To sum up, then: As methods of materials production, prescriptive technologies have brought into the real world of technology a wealth of important products that have raised living standards and increased well-being. At the same time they have created a culture of compliance. The acculturation to compliance and conformity has, in turn, accelerated the use of prescriptive technologies in administration, government, and social services. The same development has diminished resistance to the programming of people.

There are several concepts which will appear and reappear in my discussion of the real world of technology. The notion of scale is one of them. Economies of scale are intimately connected with advances in industrial production. Arguments extolling the benefits of economies of scale were as frequently heard in discussions about industrialization and the use of machines in nineteenth-century Britain as they are heard now when mergers and takeovers are debated.

Scale was a term initially used solely to indicate differences in size: It was felt that the scale of a cathedral had to be different from that of a village church; the manorhouse was built differently in scale from the cottage of a labourer. Large scale signified greater prestige, rather than improved functionality. Only when the notion of scale was applied to production technologies was an increase in scale perceived as an increase in effectiveness, and therefore as inherently beneficial to the factory owner. From being a measure of comparison, the notion of scale moved to being a figure of merit. The value-laden phrase "bigger is better" — without ever stating for *whom* it is better — comes solely out of a production-centred context.

Underlying the different uses of the concept of scale are two different models or metaphors: one is a growth model, the other a production model. Models and analogies are always needed for communication, and in order to be useful tools for discussion, models and metaphors need to be based on shared and commonly understood experiences. The features of growth, the very processes and cycles of growing, the diversity of the components of each growing organism, all have resonated through the historical written records.

Much folklore carries a prejudice against the over-grown. The giants in fairy tales are often stupid, while it's the little people who are resourceful and quick. Common experience teaches that the world is made up of things of different but appropriate sizes, and in all growth models this is acknowledged — particular sizes are appropriate for particular functioning entities or species. Implicit in any growth model is the notion that size and scale are *given* relative to any particular growing organism.

Size is a natural result of growth, but growth itself cannot be commandeered; it can only be nurtured and encouraged by providing a suitable environment. Growth occurs; it is not made. Within a growth model, all that human intervention can do is to discover the best conditions for growth and then try to meet them. In any given environment, the growing organism develops at its own rate.

A production model is different in kind. Here things are not grown but made, and made under conditions that are, at least in principle, entirely controllable. If in practice such control is not complete or completely successful, then there is an assumption, implicit in the model itself, that improvements in knowledge, design, and organization can occur so that all essential parameters will become controllable. Production, then, is predictable, while growth is not. There is something comforting in a production model — everything seems in hand, nothing is left to chance — while growth is always chancy.

Production models are perceived and constructed without links into a larger context. This allows the use of a particular model in a variety of situations. At the same time such an approach discounts and disregards all effects arising from the impact of the production activity on its surroundings. Such *externalities* are considered irrelevant to the activity itself and are therefore the business of someone else.[14] Think of a work situation, a production line. There are important factors — such as pollution or the physical and mental health of the workers — which in the production model are considered other people's problems. They are externalities.

We know today that this discounting of context and the failure to consider external and interactive effects are, in fact, a ticket to trouble. We know that the deterioration of the world's environment arose precisely from such inadequate modelling. Processes that are cheap in the marketplace are often wasteful and harmful in the larger context, and production models make it quite easy to consider contextual factors as irrelevant.

It is worthwhile stopping for a minute to see whether

we ought not to think far more in terms of growth models rather than production models, even though today production models are almost the only guides for public and private discussions. It is instructive to realize how often in the past the production model has supplanted the growth model as a guide for public and private actions, even in areas in which the growth model might have been more fruitful or appropriate. Take, for instance, education. Although we all know that a person's growth in knowledge and discernment proceeds at an individual rate, schools and universities operate according to a production model. Not only are students tested and advanced according to a strictly specified schedule (at least at the university where I have taught for the last twenty years), but the prospective university students and their parents are frequently informed that different universities produce different "products." Within all production activities, complaints of users are taken very seriously, and those complaints can often result in modifications of the production line. Thus, adverse comments from captains of industry may result at universities in the establishment of extra courses such as entrepreneurship or ethics for engineers, spelling for chemists, or fundraising for art historians. The implication is that choosing a particular university, following a particular regimen, will turn the student into a specifiable and identifiable product.

Yet all of us who teach know that the magic moment when teaching turns into learning depends on the human setting and the quality and example of the teacher — on factors that relate to a general environment of growth rather than on any design parameters set down externally. If there ever was a growth process, if there ever was a holistic process, a process that cannot be divided into rigid predetermined steps, it is education.

Similar arguments for not supplanting growth models with production models could be made in the case of health care and in many of the applications of the new biotechnologies. For me the most frightening incursions of production technologies and production thinking have happened in the new human reproductive technologies. The close monitoring of the fetus and some of the invasive prenatal technologies can only be considered as quality-control methods, with the accompanying rejection of substandard products.

On a quite different plane there is another very interesting contrast between the growth model and the production model. This situation occurs in the area of demography and population growth. You'll remember that before the Industrial Revolution there was a fascination with numbers and population increase. It was a time when people like Malthus, Ricardo, and Adam Smith were preoccupied with the growth in numbers of the lower classes.[15]

Now, oddly enough, there was no such preoccupation with the growth in numbers of the rich. Queen Victoria, for instance, had nine children. The youngest was three when the Prince Consort died. She had thirty-nine grandchildren, and none of her children or grandchildren died in infancy, as was common at the time. The drain on the public purse, one would think, of thirty-nine grandchildren of Queen Victoria was substantially larger than of thirty-nine

grandchildren of wives of miners or farmers. Nevertheless, it was the growth in numbers of the poor that
fascinated economists and statisticians. Since then
"demography" has become an area of legitimate study.

Today population forecasts are based on extensive and
reliable data. Issues related to population growth and the
resources required to sustain an increasing number of
people on earth are being discussed on the basis of reasonably factual information and developed methodology.
However, no such demographic base exists for the world's
growing population of machines and devices, as I have
stressed on earlier occasions.[16] This absence is a telling
phenomenon, since appropriate data bases could be
generated if there were the political will to do it.

The automobile, for instance, has been part of many societies for the last hundred years or so. The support
structures for the car population are in place — the production of gasoline and its delivery by service stations,
roads and bridges, car ferries, and parking garages. We
know about smog and toxic emissions, resource limitations, and transportation problems. Yet in spite of all this,
birth control for cars and trucks is not an urgent agenda
item in any public discussion. Useful statistics are hard to
come by, since nobody does the type of nose-counting for
machines and devices that is commonly applied to people.
Lots is known about the life expectancy of people in different parts of the world, about the caloric requirements for
their well-being, and so on. Almost nothing is known about
the global energy need of devices or about their lifespans.
China can embark on a rigorous one-child-per-family

policy for the sake of the country's future, and in general that policy has been approved by the world community. But where in North America, western Europe, or Japan is there serious discussion on the political level about, for instance, the need for a one-*car*-per-family policy for the sake of the country's or the world's future? Now may be the time to take machine demography seriously and enter into real discussions about *machine* population control.

The real world of technology seems to involve an inherent trust in machines and devices ("production is under control") and a basic apprehension of people ("growth is chancy, one can never be sure of the outcome"). If we do not wish to visualize people as sources of problems and machines and devices as sources of solutions, then we need to consider machines and devices as cohabitants of this earth within the limiting parameters applied to human populations.

I began this talk with a look at technology as *practice*. The common practice of particular technologies can identify people and give them their own definition; it also identifies and limits the content of what is permissible. I moved on to considerations of the division of labour and stressed the importance of prescriptive technologies. These prescriptive technologies that now encompass almost all of technological activity are, in social terms, designs for compliance, and in this I find one of the most important links between technology, society, and culture. I illustrated this with some examples from ancient China. And I touched on the concept of scale. The change in the notion of scale from a simple parameter of comparison to a parameter of merit allowed me to discuss the presence of two different models — a growth model and a production model — present throughout history. Just as prescriptive technologies have, in the real world of technology, overwhelmed holistic ones, so have production models now become almost the only pattern of guidance for public and private thought and action.

The unchallenged prevalence of the production model in the mindset and political discourse of our time, and the model's misapplication to blatantly inappropriate situations, seems to me an indication of just how far technology as practice has modified our culture. The new production-based models and metaphors are already so deeply rooted in our social and emotional fabric that it becomes almost sacrilege to question them. Thus one may question the value of people (to go back to the issues of human demography I just mentioned), but not the fundamental value of technologies and their products.

But question we must. It is my view that today's real world of technology is planned and run on the basis of a production model that is no longer appropriate for the tasks that we want to undertake. Any critique or assessment of the real world of technology should therefore involve serious questioning of the underlying structures of our models, and through them, of our thoughts. To quote Kenneth Boulding again:

> We cannot walk before we toddle,
> but we may toddle much too long
> if we embrace a lovely Model
> that's consistent, clear and wrong.[17]

II

I called these lectures *The Real World of Technology* for two reasons. One is that I wanted to pay tribute to C. B. Macpherson and his 1965 Massey Lectures, "The Real World of Democracy." I intend to look at technology the way Macpherson looked at democracy, as ideas and dreams, as practices and procedures, as hopes and myths. The second reason is that I wanted to discuss technology in terms of living and working in the real world and what this means to people all over the globe. This is the "real" part in the title, and this is what I wish to address now, having spent some time in the first talk on certain aspects of technology as practice.

When I talk about reality, I'm not trying to be a philosopher. I think of reality as the experience of ordinary people in everyday life. There are different levels of reality, and I would like to go quickly through these levels and then look at how they are influenced by the technologies around which our real world is built.

The first level of reality is that nitty-gritty stuff, the direct action and immediate experience, the sort of thing I like to call *vernacular reality*.[1] It's bread and butter, soup, work, clothing and shelter, the reality of everyday life. This reality is both private and personal, but it is also common and political. Feminists have often stressed that the personal is political,[2] and it is this realization that has affected much of my own thinking; it will also permeate through what I am going to say here.

For the purposes of these lectures, I will call *extended reality* that body of knowledge and emotions we acquire that is based on the experience of others. Here we have all those experiences that we would have had, had we been there. These are the experiences of war, of depression, of old age, of foreign travel, of religious experience that those who were gifted enough to put into words have told us about. The extended reality includes also artifacts — that's the stuff we collect in museums — which we try to make part of our own reality because we like to draw on that continuity, on that experience of the past, so much so that some of us prefer to hear old music performed on original instruments in order to make the linkage to another time more concrete.

Over and above this, we live with what I call *constructed* or *reconstructed reality*. Its manifestations range from what comes to us through works of fiction to the daily barrage of advertising and propaganda. It encompasses descriptions and interpretations of those situations that are considered archetypal rather than representative. These descriptions furnish us with patterns of behaviour. We consider these patterns real, even if we know the situations have been constructed in order to make a particular pattern very clear and evident. So when we read Dostoyevsky, we know that the Grand Inquisitor is not just an episode in Russian history, it's a pattern of inquisition, the prototype of what happens to the powerless in front of the powerful all over the world. Every Christmas, Dickens' Scrooge is paraded as the archetype of grumpy selfishness. The constructed or reconstructed realities are part of the fabric that holds the common culture together. They become so much a part of the vernacular reality that a newcomer confronts with puzzlement references that just cannot be figured out. This happened to me a great deal when I first came to Canada, and it is as awkward as hearing people laugh at a joke and not understanding what is funny about it.

Finally there is *projected reality* — the vernacular reality of the future. It is influenced or even caused by actions in the present. Heaven and hell or life after death were — and are still — for some people projected reality. Today there's also "the future" itself, the five-year plan, the business cycle — and these can influence people's actions and attitudes as much as or more than the price of bread or the level of wages.

All levels of reality have been profoundly affected by science and technology, but before illustrating this particularly with respect to the realities of time and space, I need to touch briefly on two other aspects of the real world of technology: one is the relationship between science and technology; the other is the nature of experience.

With respect to the relationship between science and technology, it has often been assumed that science is a prerequisite for technology. I'm not sure whether this has ever strictly been true. Certainly in seventeenth- and eighteenth-century western Europe science did stimulate a large number of technologies. However, today there is no hierarchical relationship between science and technology. Science is not the mother of technology. Science and technology today have parallel or side-by-side relationships; they stimulate and utilize each other. It is more appropriate to regard science and technology as one enterprise with a spectrum of interconnected activity than to think of two fields of endeavour — science as one, and applied science and technology as the other. Thus when I speak of modern science and technology, I mean this unit of enterprise until I specify other constraints.

In spite of what I've just said I want to speak for a moment about the scientific method. Science as well as technology is, after all, more than just a body of knowledge; it is a set of practices and methods. The scientific method as we understand it in the West is a way of separating knowledge from experience. It is the strength of the scientific method that it provides a way to derive the general from the particular and then, in turn, allows general rules and laws to be applied to particular questions. Consequently, somebody can today go to a university and learn how to build bridges from somebody who has never built a bridge.[3]

The scientific method works best in circumstances in which the system studied can be truly isolated from its general context. This is why its first triumphs came in the study of astronomy.

On the other hand, the application of the general to the specific has been much less successful in situations where generalization was achieved only by omitting essential considerations of context. These questions of reductionism, of loss of context, and of cultural biases are cited quite frequently by critics of the scientific method.[4] We hear much less about the human and social effects of the separation of knowledge from experience that is inherent in any scientific approach. These effects are quite widespread and I think they can be serious and debilitating from a human point of view.

Today scientific constructs have become *the* model of describing reality rather than *one* of the ways of describing life around us. As a consequence there has been a very marked decrease in the reliance of people on their own experience and their own senses. The human senses of sight and sound, of smell and taste and touch, are superb instruments. All senses, including the so aptly named "common sense," are perfectible and it's a great pity that we have so little trust in them. For instance, people know at what point an ongoing noise will give them a headache, but all too often they feel the need for an expert with a device that measures the noise in decibels. The expert then has to compare the noise level measured with a chart that indicates the effect of noise levels on the nervous system. Only when that chart and the expert say, "Yes, indeed, the noise level is above the scientifically established tolerance range," do people believe that it was indeed the noise and not a figment of their imagination that gave them persistent headaches. I'm not talking here about an either-or situation in which either personal experience or an established measuring procedure is paramount; what I am talking about is the downgrading and the discounting of personal experience by ordinary people who are perfectly well equipped to interpret what their senses tell them. I dwell on this because the downgrading of experience and the glorification of expertise is a very significant feature of the real world of technology.[5] Sometimes it is important to stress that because the scientific method separates knowledge from experience it may be necessary in case of discrepancies to question the scientific results or the expert opinion rather than to question and discount the experience. It should be the experience that leads to a modification of knowledge, rather than abstract knowledge forcing people to perceive their experience as being unreal or wrong.

Feminist authors in particular have often called for changes in the way in which the social and human impact of technology is evaluated.[6] They have stressed the need to base such evaluation on the experience of those who are at the receiving end of the technology. They have also drawn attention to the overbearing role of experts in the lives of those who, like many women, have no claim to certified expertise because most of their knowledge is not separated from their experience. I foresee great changes in the evaluation of technology and almost all of them will come from bringing in direct experience, which is, after all, the central core of vernacular reality.

All the realities I mentioned, the vernacular and the extended, the constructed and the projected, have been profoundly affected and distorted by modern technology. We ought to keep in mind that the effects we perceive as so large come from technologies that are very recent in historical terms. For example, practical applications of electromagnetic and electronic technologies that have so profoundly changed the realities of the world are not more than one hundred and fifty years old. Think for a moment of the speed of transmission of messages; this really didn't change between the time of antiquity and about 1800. Whether Napoleon or Alexander the Great, even emperors had to rely on teams of horses and riders to send their messages and receive their responses. Then suddenly, around 1800, the speed of transmission of messages changed from the speed of galloping horses to the speed of the transmission of electricity — the speed of light. Before 1800 optical signals were exchanged, particularly in the military. But it was really only around 1800 that batteries were developed sufficiently and long-distance transmission of electrical signals became feasible, and it was 1825 before adequate electromagnets produced currents and fields that were practically useful.

In 1833, Gauss and Weber, two German professors, strung a mile and a half of copper wire over the roofs of Göttingen and sent electrical impulses along it. And at that time in the United States, Samuel Morse (of the Morse Code) experimented with signal transmission. It was 1844 before he was able to string a line over sixty kilometres and transmit a message in Morse Code. He transmitted the sentence, "What hath God wrought?" and thus began the first useable method for the quick transmission of messages. In 1876, Alexander Graham Bell received his patent. Faraday conducted his crucial experiments in 1833, Marconi experimented during the beginning of the first decade of this century, and people began to play with radios only after the First World War. So it was essentially during the last 150 years that the speed of transmission of messages truly changed. This, in turn, so completely changed the real world of technology that we now live in a world that is *fundamentally* different.

In terms of the realities that we have discussed, the message-transmission technologies have created a host of pseudorealities based on images that are constructed, staged, selected, and instantaneously transmitted. I'm talking here about the world of radio, television, film, and video. The images create new realities with intense emotional components. In the spectators they induce a sense of "being there," of being in some sense a participant rather than an observer. There is a powerful illusion of presence in places and on occasions where the spectators, in fact, are not and have never been. Edward R. Murrow's phrase, "You are there," led his audience to believe that they were somehow "present" at important international events.

In French the news is called *les actualités*, although there is very little that is actual and real in the images and the stories that we see and hear. The technological process of

Image-making and image transformation is a very selective one. It creates for the eye and ear a "rendition" rather than an "*actualité*." Yet for people all around the world the image of what is going on, of what is important, is primarily shaped by the pseudorealities of images. The selective fragments that become a story on radio and television are chosen to highlight particular events. The selection is usually intended to attract and to retain the attention of an audience. Consequently, the unusual has preference over the usual. The far away that cannot be assessed through experience has preference over the near that can be experienced directly. There is a sense of occasion that is conducive to making what is seen to appear seem as if it was *all* that happened. Anyone who has ever been at a demonstration and then seen their own experience played back on television knows what I mean. Frequently a small counter-demonstration to a large demonstration is treated as if it were the main event. Side-shows move into the centre and the central issues become peripheral.

Because there are now technical and economic means of creating these pseudorealities of images and exposing a large part of the globe to them, experienced realities and their dynamics have changed, and keep on changing.[7] Today the question of whether or not an event is reported and televised may be more important than the content of the event itself. The presence of reporters, camera crews, or external observers affects events as they take place, sometimes initiating new actions. Think of the television cameras in China, South Africa, or eastern Europe. Think about the PCBs in Quebec. Among all the difficult problems of disposing of toxic waste, the destruction of PCBs is a relatively minor and doable task if you compare it to, for instance, the problem of nuclear-waste disposal. Nevertheless, the political dimensions of the PCB issue took a quantum jump the moment the events entered the world of images.

The encroachment of pseudorealities and images into the real world of experience of ordinary people occurs all over the world, but there are important *local* applications and implications. I only have to remind you, for instance, of the effect of images on candidates for political office. A good, warm image doesn't say anything about competence or integrity. Still, image considerations loom very large in terms of political advancement and success, because political responses, more often than not, are now based on images. And what seems extraordinary to me is that these media images have so permeated every facet of life that they are no longer perceived as external intrusions or as pseudorealities except by media professionals, and only professionals and academics discuss these images. There is no common discourse about how the images were formed, how they were gathered, how they got into our living rooms.

Media images seem to have a position of authority that is comparable to the authority that religious teaching used to have. The images seem infallible in the way the Pope's authority was unquestionable prior to the Reformation. In the real world of technology, I think, we would be well advised to question the authority of the images in the manner in which the Reformation questioned the authority of the Pope. And I say this because throughout the Middle Ages the Church, with its doctrine and its religious teaching, was the authority that prescribed the conduct of social and political relations. The Reformation challenged that authority of the Church to be the sole arbiter of individual conduct beyond the individual's own conscience and discernment. Today, technological rationales have very much the force and authority of religious doctrine, including the notion that the laity is unfit to question doctrinal content and practice. It is in a spirit of questioning authority that we should ask, "What about people who are at the receiving end of technologically produced pseudorealities of images?" Their work has changed as their lives have changed. Life and work have been restaged by external forces. The literature of television and advertising is testimony to that, but more so is the practice of both. The reconstructed world of images has taken over much of our vernacular reality, like an occupation force of immense power. And somewhere, someone will have to ask, "How come the right to change our mental environment — to change the constructs of our minds and the sounds around us — seems to have been given away without anybody's consent?"

While it's possible, in theory, to opt out of this world of images, in practice one can really only do that in a very limited way. Of course I don't have to watch television, I don't have to listen to the news, and many of us have indeed begun to cultivate other channels of information that are more directly related to the life experiences of our contemporaries. But the pseudorealities and the images are there, and the world is structured to believe in them.

If I want to promote change I need to understand and appreciate the structuring of the images, even if I don't trust their content. Opting out by individuals really doesn't change the agenda of what is urgent and what is not, unless there is a collective effort to supplement and substitute the images with genuine experience. Just because the imaging technology has emphasized the far over the near, the near doesn't go away. Even though the abnormal is given a great deal more play than the normal, the normal still exists and, with it, all its problems and challenges. But somehow observing a homeless person sleeping in the park around the corner doesn't seem to register as an event when it's crowded out in the observer's mind by images from far-away places.

Let me give you a recent example. Canadian news has been full of the events in eastern Europe. We have seen East Germans going to the West; we've seen the events in Poland. Resonating in the reports was a note of cheering joy: "Yes, democracy means so much to them; they struggle for it; they win; we cheer." At the same time very significant events in our own country — such as the cuts to Via Rail — are not discussed in Parliament, not voted upon after a democratic debate, but decided in seclusion by Cabinet. So you can understand why I feel that danger lurks when the far so outperforms the near. As a community we should look at what the new technologies of message-forming and -transmitting do to our own real world of technology and democracy. This is why I have a sense of urgency to map the real world of technology, so that we might see how in our social imagination the near is disadvantaged over the far. We should also understand that this does not have to be so.

The strong impact of the world of images on people's reality has yet another component. Viewing or listening to television, radio, or videos is *shared experience carried out in private*. The printing technologies were the first ones that allowed people to take in separately the same information and then discuss it together. Prior to that, people who wanted to share an experience had to be together in the same place — to see a pageant, to listen to a speech. Then, printed text — quoted and requoted — yielded some of the common information. Now there are new, high-impact technologies and these produce largely ephemeral images. The images create a pseudocommunity, the community of those who have seen and heard what they perceive to be the same event that others, who happened not to have watched or listened, missed for good. Just listen to a discussion of a hockey game — or, for that matter, to a discussion of a leader's debate — that no one present attended. The talk proceeds as if all had been there. In this manner, pseudorealities create pseudocommunities.

There are times when the pseudocommunities of viewers and listeners spawn real communities of common concern. Many credit the formation of communities of opposition to the Vietnam War and their political effectiveness to television images of the war. One can perceive these communities of concern as originating from within the pseudocommunities of viewers and listeners. A similar process brought together those who have acted jointly to help victims of famine in Africa — Live-Aid, the Space Bridge movement — and formed international groups that focus attention on particular environmental issues. Since normally only a fraction of the pseudo-community become members of the real and active community, the possibility of forming such groups may be greater in the case of broadly based international concerns that are "the far" for most viewers than in the case of specific problems of "the near."

Nevertheless, this reforming of community is a very hopeful sign, particularly against the background of passivity that listening and viewing commonly entails and which, I need to stress again, is so reinforced by the avoidance of direct experience. However, the reformed communities seem to be more successful in addressing manifestations than in addressing the root causes of the concerns that brought them together. Temporary relief from hunger does not eliminate famine, and withdrawal of U.S. troops from Vietnam did not address the fundamental reasons for aggression.

To recap, technology has developed practical means to overcome the limitations of distance and time. Devices, organizations, and structures have been created for this purpose and they are now an integral part of our social and political landscape. Everyone's vernacular reality has changed. In addition to carrying out established tasks in a different manner (for instance, not to write but to phone, not to use the mail or courier but to fax), there are genuinely new activities that are possible now that could not have been done without the new technologies and their infrastructure. They are, in the main, related to the transfer, storage, and reconstruction of information. Some of these affect our approaches to and perceptions of the future, that is, the projected realities.

The technological possibilities for information gathering, storage, and evaluation, interwoven with a tight net of administrative infrastructures, have made it possible to treat certain parts of the future as parts of the present. Let me give you one illustration. It's a useful, but not unique, example. There are futures markets and commodity exchanges, and there is such a thing as "trading in futures." This means that one can buy or sell shares of crops that have not yet been grown, that one is speculating on the price of animals that are not yet born or on the demand for products not yet made. Still, such trading activities are not mere figments of the imagination; they are not totally removed from reality just because the crops have not yet grown, because the *results* of those trading activities are real. People make money or lose money in futures trading; rent is paid from transactions made on dealings that are essentially hypothetical. Thus money ties the present and the future together in a way that did not exist in an earlier world. What is new, as so often is the case when new technical means restructure social and economic activities, is the organized, standardized, "normal" appearance of transactions that are to be carried out in the future; the deals fit without discontinuity into the procedures for transactions entirely limited to the present.

The future is thus perceived and handled as a structural and technical extension of the present.

There is a lot of talk about global crises and "our common future."[8] However, there is far too little discussion of the *structuring* of the future which global applications of modern technologies carry in their wake. What ought to be of central concern in considering our common future are the aspects of technological structuring that will inhibit or prevent future changes in social and political relations.

And now I'd like to focus for a moment on the human consequences which are particularly evident in what are called the communications technologies, and which I would like to call the "non-communications" technologies because very often that word, "communication," is a misnomer. Whenever human activities incorporate machines or rigidly prescribed procedures, the modes of human interaction change. In general, technical arrangements reduce or eliminate *reciprocity*. Reciprocity is some manner of interactive give and take, a genuine communication among interacting parties. For example, a face-to-face discussion or a transaction between people needs to be started, carried out, and terminated with a certain amount of reciprocity. Once technical devices are interposed, they allow a physical distance between the parties. The give and take — that is, the reciprocity — is distorted, reduced, or even eliminated.

I am very fond of a Ben Wicks cartoon which illustrates this point beautifully. The cartoon shows a repairman in a living room removing a television set with a smashed screen. Next to the set stands a man on crutches, one foot heavily bandaged, to whom the repairman says, "Next time Trudeau speaks, just turn the set off." I think that says it all. It says that a personal response of the kind that the man in the cartoon was obviously eager to give can neither be given nor received when communication is mediated by technology. Any reciprocity is ruled out by design. This loss of reciprocity is a continuing form of technologically executed inequality. It has very profound political and psychological consequences.

I'd like to stress that reciprocity is not feedback. Feedback is a particular technique of systems adjustment. It is designed to improve a specific performance. The performance need not be mechanical or carried out by devices, but the purpose of feedback is to make the thing work. Feedback normally exists within a given design. It can improve the performance but it cannot alter its thrust or the design. Reciprocity, on the other hand, is situationally based. It's a response to a given situation. It is neither designed into the system nor is it predictable. Reciprocal responses may indeed alter initial assumptions. They can lead to negotiations, to give and take, to adjustment, and they may result in new and unforeseen developments.

I emphasized earlier the extent to which the new technologies of image procurement have invaded the real world of technology. By design, these technologies have no room for reciprocity. There is no place for response. One may want to speculate for a moment whether this technological exclusion of response plays a part in the increasing public acceptance of the depiction of violence and cruelty. I find it hard to imagine anyone actually standing next to a person who is being hurt or abused and enjoying the sight and sound of the experience, nor can I imagine such a direct observer not intervening or at least feeling guilty for having failed to do so. On the other hand, violence depicted on a screen appears to be acceptable and entertaining. It doesn't seem to matter how violence is depicted or how degrading or obscene it is. Viewers are not called to respond. They are not "there," where the action is.

The notion of reciprocity may also help to explain an apparent contradiction between responses to viewing or listening in particular settings. Have you ever been in an overflow audience that had to listen to a lecture on closed-circuit television in an adjacent room because the main auditorium was full? Most people in this situation are quite disgruntled and feel cheated, although they see and hear exactly the same thing as people in the main auditorium. Still, they say, it doesn't feet real, it doesn't feel right. By the same token, the great hope of using TV or film to let many students benefit from exceptional interpreters, letting them be part of an outstanding lecture, has not been realized. Students just do not like to be taught by a television screen. This does not mean that one can never make good use of a new technology, but here it seems to work only in a supplementary mode, in terms of illustration and extension. Teaching on television is rarely engaging in the way news reports of flood and famine are. The pseudoimages and events reported by regular television are dramatic renditions, carefully selected and assembled fragments. This is different from the camera or the tape recorder in the lecture room, recording everything without selection or comment.

On the other hand, viewers are rivetted to televised proceedings of judicial inquiries or public hearings. It seems that in situations where reciprocity is neither permissible nor desired — such as when observing an actual inquiry — images are acceptable substitutes for reality. However, whenever the potential for reciprocity exists and is valued — as in the lecture or teaching situation — images won't do.

We should reflect on the possibility that technology that produces pseudorealities of ephemeral images and eliminates reciprocity also diminishes the sense of common humanity. This may sound dramatic, but such a development can start with very simple but pervasive steps. Where there is no reciprocity, there is no need for listening. There is then no need to understand or accommodate. For kids this can mean that one doesn't have to be moderately civil to one's younger sister because she is the only one to play with; television allows entertainment without the cooperation of anybody. In school, there is no argument or negotiation with the computer. Sharing work among students takes on a different meaning. Women who work in automated offices often report how much human isolation the automation has brought for them. When work isn't shared, the instruments of cooperation — listening, taking note, adjusting — atrophy like muscles that are no longer in use.

One illustration of technologically induced human isolation: When I go to work in the morning I often meet a neighbour and her ten-year-old daughter. Every day they walk side by side to the bus stop, each plugged into her own Walkman, isolated from one another and from the rest of the world. Such is the real world of technology. The question that lingers on in my mind is this: How will our society cope with its problems when more and more people live in technologically induced human isolation?

Let me emphasize again that technologies need not be used the way we use them today. It is not a question of either no technology or putting up with the current ones. Just remember that even in the universe of constructed images and pseudorealities there still exists a particular enclave of personal directness and immediacy: the world of the ham-radio operator. It is personal, reciprocal, direct, affordable — all that imaging technology is not — and it has become in many cases a very exceptional early warning system of disasters. It is a dependable and resilient source of genuine communication. I am citing this example so as not to leave the impression that the technological reduction of meaningful human contact and reciprocal response is inherently inevitable.

In this lecture I have tried to demonstrate the changes that technology has brought to our perceptions of reality. Many of these changes have become such an integral part of the fabric of existence in the real world of technology that the resulting distortions of human and social relations are now considered normal and beyond questioning. I hope that I have also shown how one can question and,

I hope, mitigate the encroachments of pseudorealities.

In the next lecture I will focus on technology as a catalyst for the spread of control and management, and on the infrastructures that have become part of our society because of this expansion.

III

During the first lecture in this series, I showed how the growth of prescriptive technologies provided a seed-bed for a culture of compliance. I contrasted holistic and prescriptive technologies and stressed the differences in their respective divisions of labour. It is characteristic of prescriptive technologies that they require external management, control, and planning. They reduce workers' skill and autonomy. But they are exceedingly effective in terms of invention and production.

In the second lecture I indicated how much the realities within which today's real world of technology exist have changed from those of the past. Modern realities have been restructured through technological activities. This development has affected human and political relations. The communications technologies, some of which

I would rather call non-communications technologies, provided us with illustrations for the concept of reciprocity, and the ways in which advances in certain technologies have been accompanied by genuine losses of reciprocity.

Now I would like to turn to technology as a catalyst for the spread of control and management.

The fact that citizens are more and more stringently controlled and managed is often considered as normal and fundamentally beyond questioning, as a necessary feature of technological societies. Technology has been the catalyst for dramatic changes, in the locus of power.

Traditional notions about the role and task of government, for instance, or about what is private and what is public, are in the light of these changes more often akin to fairy tales than to factual accounts of possible relationships of power and accountability.

Shifts of power and control are going on all the time. These processes are complex and interactive. Technologies, as I have stressed throughout these lectures, exist in particular contexts, and these contexts are usually fluid and changeable. Within a given context, the relationship between tool and task is of fundamental importance. Historians have often pointed out that when special tools become available to carry out particular tasks the success of these tasks will, by necessity, encourage the further use of the tools, which may then be improved and adapted to other tasks. But the success and spread of a particular tool — and this tool can be organizational or administrative as well as mechanical — has another consequence. Any task tends to be structured by the available tools. It can appear that the available tools represent the best or even the only way to deal with a situation. This happens every day. If you have a particular type of kitchen equipment, you begin to slice and dice as you have never done before. Other means of food preparation become less attractive and you may eventually forget about them. If your lab gets an electron microscope, you will find it difficult to persuade students to use optical microscopy.

Tools often redefine a problem. Think, for instance, of speeding and radar traps. Let's go back to the purpose of speed limits. They were instituted to enhance safety, not to produce criminality. One way of enforcing speed limits used to be the judicious presence of clearly marked police cruisers on our highways. The police drove at the speed limit and by this tactic brought the traffic pattern into compliance with the regulations. The tool of radar traps brought another dimension into the situation. The emphasis shifted from common safety to individual "deterrence." It was felt that the fear of being caught and fined would be a better way of enforcing the regulations. Next came a technological option of avoiding the radar trap, using what's commonly called a "fuzz-buster." Now the motorist, concerned less with safety than with criminality, buys an avoidance device, whether it is outlawed or not. The next player in the speeding game is a device for law-enforcement officers to detect the presence of a fuzz-buster. And now there seems to be a new generation of widget on the horizon which those with a fuzz-buster can use to detect the counter-technology of law enforcement. And so it goes.

The common problem of road safety has been transformed into the private problem of fines and demerit points and into a technological cat-and-mouse game. One might say that the technological tools designed to establish random criminality have prevented the development of techniques to establish collectively safe driving patterns. Thus it may be wise, when communities are faced with new technological solutions to existing problems, to ask what these techniques may *prevent* and not only to check what the techniques promise to *do*.

The real world of technology is a very complex system. And nothing in my survey or its highlights should be interpreted as *technological determinism* or as a belief in the autonomy of technology *per se*. What needs to be emphasized is that technologies are developed and used within a particular social, economic, and political context.[1] They arise out of a social structure, they are grafted on to it, and they may reinforce it or destroy it, often in ways that are neither foreseen nor foreseeable. In this complex world neither the option that "everything is possible" nor the option that "everything is preordained" exists.

Those who study complexity as a subject of research and enquiry have often concerned themselves with technology as a complex system. In 1985 the U.N. University called a conference on the science and praxis of complexity.[2] Jacques Ellul, the French scholar and noted critic of technology, called one of his books *The Technological System*.[3] Ellul's analysis, as well as the more mathematical papers in the U.N. conference, make it very clear how tightly interlinked the developments of technology can be. A change in one facet of technology, for instance the introduction of computers in one sector, changes the practice of technology in all sectors. Such is the nature of systems.[4]

Personally, I much prefer to think in terms not of systems but of a web of interactions. This allows me to see how stresses on one thread affect all others. The image also acknowledges the inherent strength of a web and recognizes the existence of patterns and designs. Anyone who has ever woven or knitted knows that one can change patterns, but only at particular points and only in a particular way so as not to destroy the fabric itself. When women writers speak about reweaving the web of life,[5] they mean exactly this kind of pattern change. Not only do they know that such changes can be achieved but, more importantly, they know *there are other patterns*. The web of technology can indeed be woven differently, but even to discuss such intentional changes of pattern requires an examination of the features of the current pattern and an understanding of the origins and the purpose of the present design.

So let me now turn to questions of technologically induced changes in social and political patterns and to the subject of planning. It is normal for any society to evolve social institutions and to structure its social activities so that the power and control of the structuring authority is maintained and advanced. As I pointed out in the first lecture, prescriptive technologies have historically been fine instruments for such structuring; such technologies existed well before the introduction of machines.

Here the work of Michel Foucault, the great French social historian, is of major importance. I'm referring particularly to his book *Discipline and Punish*, published in 1975, in which he develops a thorough analysis of the social history of the seventeenth and eighteenth centuries, particularly of the changing structures of French schools and hospitals, military institutions and prisons. He shows how, in the beginning of the eighteenth century, a new notion of discipline enters the secular sphere. Discipline, well established in monastic communities, begins to be applied in great detail within secular groups; with it comes detailed hierarchical structures, rank and sub-rank, drill, surveillance, and record-keeping.

At the time, in the 1740s, a very influential book was published by La Mettrie called *L'Homme-machine*, which means "man-the-machine." Its publication symbolized the changes that were occurring in French society. La Mettrie looked upon the human body as an intricate machine, a machine that could be understood, controlled, and used. Foucault points out that the discovery of the body as object and instrument of power led to a host of regimes of control for the efficient operations of these bodies, whether they were the efficiencies of movement, the measured intervals of the organization of physical activities, or the careful analysis and timing of the tasks bodies could perform, usually in unison. Foucault reminds us that in the course of the seventeenth and eighteenth centuries the disciplines of exercise, training, and work became, in general, the prescriptions for domination.

Let me quote directly from Foucault.

> The human body was entering a machinery of power
> that explores it, breaks it down and rearranges it. A
> "political anatomy" was being born . . . it defined how
> one may have a hold over others' bodies, not only so
> that they may do what one wishes, but so that they
> may operate as one wishes, with the techniques, the
> speed and the efficiency that one determines. Thus
> discipline produces subjected and practised bodies,
> "docile" bodies.[6]

These developments resulted in schools where the activities of the pupils were timed to the minute. They resulted in the analysis of movements of soldiers, as in the 1743 regulations which specified six steps to bring the weapon to the soldier's foot, four to extend it, thirteen to raise it to the shoulder, and so on. The drill based on these movements and steps made the military unit into a machine at the command of its superior. In the same manner the great workshops of France, prior to the Industrial Revolution, already had a detailed labour discipline that corresponded quite closely to that of the army.

Foucault recounts how French towns were managed during the Plague — how the city areas were sub-divided, how all were centrally controlled, and how total discipline was enforced by threat of execution. In a milder vein, he shows designs for compliance embedded in the architecture of prisons, hospitals, and training schools. Their structural arrangements were later incorporated into the designs of factories.

It was into this socially and politically well prepared

soil that the seeds of the Industrial Revolution fell. The factory system, with its mechanical devices and machines, only augmented the patterns of control. The machinery did not create them. The new patterns, with their minute description of detail, their divisions of labour, and their breakdown of processes into small prescriptive steps, extended quickly from manufacturing into commercial, administrative, and political areas. In England and in France, changes quickly occurred in institutions as diverse as banks and prisons. Planning thrived as an activity closely and intimately associated with the exercise of control.

In Britain, the Industrial Revolution led to developments similar to those described by Foucault. They were considerably accelerated when machines entered into the workplace. To plan with and for technology became the Industrial Revolution's strongest dream. The totally automated factory — that is, a factory completely without workers — was discussed by Babbage and his contemporaries in the early nineteenth century.[7] It took, though, another two hundred years of technical development and the creation of new infrastructures to make such schemes a practical feature of the real world of technology.

I began this lecture by reiterating that technology exists in context. Patterns laid down in the practice of technology become part of a society's life. Most new patterns are elaborations and continuations of older patterns and so I find it helpful to trace current patterns and find historical roots. You may ask, why go back so far into history when I keep saying the world has so fundamentally changed?

The reason for dwelling on historical situations is that patterns do have very deep and profound roots. It is my conviction that we are at the end of a historical period in which processes and approaches that initially had been exceedingly constructive and helpful have run their course and are now in many ways counterproductive. The roots of some of the most important patterns that have shaped the real world of technology go back to seventeenth- and eighteenth-century Europe.

The emergence of new social patterns in the eighteenth century offers an understanding of how the massive changes of the Industrial Revolution could have taken place in such a short time and with relatively little social upheaval. In turn, the change in the structure of society and the nature and organization of work and production during the Industrial Revolution became a pattern onto which our real world of technology with its much more extended and sophisticated restructuring is grafted. It is for that reason that I want to spend a little more time on events that occurred during the Industrial Revolution.

While the eighteenth century exercised control and domination by regarding human bodies as machines, the nineteenth century began to use machines alone as instruments of control. For the British manufacturers, machines appeared more predictable and controllable than workers. The owners of factories dreamt of a totally controlled work environment, preferably without any workers. If and where workers were still needed, they were to be occupied with tasks that were paced and controlled by machines.

Industrial layout and design was often more a case

of planning *against* undesirable or unpredictable interventions than it was of planning *for* greater and more predictable output and profit. Thus the resistance of workers to mechanization and to the introduction of machines into the process of their own work must be understood, as David Noble in his essay "Present tense technology" points out,[8] in terms of a clearly perceived loss of workers' control and autonomy. It was not resistance to technology *per se* so much as an opposition to the division of labour and loss of autonomy that motivated the workers' resistance. (In fact, there were many interesting inventions by the very people who most strongly opposed the introduction of machinery into the workplace.) What the Luddites and other groups of the period clearly perceived was the difference between work-related and control-related technologies.

At the core of much political debate at the time lay the question of the division of labour and the introduction of machinery into the workplace. There was extensive parliamentary and public discussion about whether the replacement of workers by machines was right and proper and whether this process of mechanization and industrialization was the only way to prosperity. The description of life in the new industrial Britain by Dickens and other Victorian writers should be regarded as one form of this public discourse.

Already at the beginning of the seventeenth century, the interest in the division of labour and the efficiency of production was quite intense in Britain. In the 1680s, Sir William Petty, in a report related to the growth of the city of London, wrote in favour of the division of labour as follows: "In the making of a *Watch*, If one Man shall make the *Wheels*, another the *Spring*, another shall Engrave the *Dial-plate*, and another shall make the *Cases*, then the *Watch* will be better and cheaper, than if the whole Work be put upon any one Man."[9] Prescriptive divisions of labour were acknowledged to be a mode of production that led to better and cheaper products. Yet the step of going from the artisan-type of holistic technology to the production of greater volumes based on a division of labour as described by Petty was not easy. It required larger factories, greater capital, and another type of organization of both work and workers.

These developments took time and it was a hundred years later that the mills of England — the weaving shops as well as the heavy industries — began to use machinery and reach levels of production that allowed a prescriptive division of labour. The new steam engines offered the possibility of running much larger installations and, by the 1820s, workers really began to see their activities change. Parliamentary and political discussion of industrial problems became an important part of Britain's public life. Was it morally right that, in the name of trade, prosperity, and efficiency, the mode of work could change so drastically that many people became uprooted and deprived of their livelihood?

Maxine Berg, a British social historian, has documented the context of the debates of the "machinery question" in her remarkable book of that name.[10] She reminds us that the human costs of mechanization and machine-dictated divisions of labour were well recognized. Lord Byron, for instance, gave a passionate speech in the House of Lords on behalf of the weavers who resisted the new technological arrangements. There were interesting proposals for alternatives, particularly the plans by Robert Owen and his associates.[11] They addressed the concerns of control in the workplace through organizational innovations such as profit-sharing among workers and cooperative labour practices. Owen also dealt with what he called "spade husbandry" (we would call it organic farming), since farm machinery was introduced at the same time and was part of the same political debate. Many of Owen's arguments resonate again today in debates on the social and environmental impact of technology. In nineteenth-century Britain, Owen's alternatives were not taken seriously.

There is another commonality between our time and the period of the Industrial Revolution: Both ages had irrationally high expectations of the beneficial effects of science and technology voiced by their respective proponents. Machines — or today, electronic devices — were soon to overcome the physical and mental shortcomings of mere humans. Machines, after all, did not drink; they did not require moral guidance of the kind Victorians lavished on the working class. Today's banking machines do not belong to a union or need maternity leave — although even the most clever computers can catch amazing viruses.

There was, however, one noticeable blind spot in the contemporary considerations of mechanization during the Industrial Revolution that is not part of the repertoire of modern technological expansionists. The proponents of technology in the 1840s were very enthusiastic about replacing workers with machines. But somehow I find no indication that they realized that while production could be carried out with few workers and still run to high outputs, *buyers* would be needed for these outputs. The realization that though the need for workers decreased, the need for purchasers could increase, did not seem to be part of the discourse on the machinery question. Since then, however, technology and its promoters have had to create a social institution — *the consumer* — in order to deal with the increasingly tricky problem that machines can *produce*, but it is usually people who consume.

Usually, but not always. Today machines and institutions have themselves become major consumers of industrial production. Cars, for instance, now need far more devices in them to run than were required ten years ago. Administration and accounting have opened markets for industrial productions that now serve another layer of prescriptive technologies.

Nevertheless, there are a lot of common threads in the patterns of the Industrial Revolution and in the pattern of the real world of technology. One thread that is prominent, strong, and lasting was beautifully identified by Lewis Mumford in 1944 when he wrote in *The Condition of Man*, "Between the 13th and the 19th century, one may sum up the changes in the moral climate by saying that the seven deadly sins became the seven cardinal virtues. Avarice ceased to be a sin. The minute attention to the care of worldly goods, the hoarding of pennies, the unwillingness to spend one's surplus on others, all these habits were useful for capital saving. Greed, gluttony, avarice, envy and luxury were constant incentives to industry."[12]

In Britain, Robert Owen's road of cooperation and profit-sharing was not taken. Industrialization, which constituted in Robert Owen's time mere islands of new organizational patterns in a sea of traditional life, proceeded rapidly. In the hands and minds of the people of Europe and North America production-patterned activities and experiences began to crowd out growth-patterned ones. Within this historical development, several facets are of particular interest in our discussion of the real world of technology. The facet I would now like to address relates to the need for large-scale public infrastructures in order to develop and sustain industrial enterprises.

The argument that I would like to put before you is the following: Since the time of the Industrial Revolution, the growth and development of technology has required as a necessary prerequisite a support relationship from governments and public institutions that did not exist in earlier times. Here one wants to consider in the first place the technologies of transport — moving raw materials and final products along the road, rail, air, and water, as well as moving energy and information from points of generation to points of use. Such distribution systems and their technologies link the so-called private spheres of industry and commerce to the public spheres of local and central governments. In order to operate successfully, the industrial production technologies require permanent transportation and distribution structures. In all countries the public sphere has supplied these infrastructures and has adjusted itself accordingly. Arranging to provide such infrastructures has become a normal and legitimate function of all governments.

The infrastructures for the distribution of electricity are a case in point. As a commodity, electricity has to be generated more or less continually and it cannot be stored in the way coal or coke can be piled up. Furthermore, those products of industry that need electricity to work can only be sold and used where dependable supplies of electricity are at hand. Thus, when electrical technologies entered the scene, planning and providing for industry became an increasingly important part of the activities of governments. These were activities for the development and expansion of technological enterprises, although in terms of public policy they have always been interpreted as public activities in the public interest.

Take Ontario Hydro as one example. Sir Adam Beck saw it as one of the major obligations of the Government of Ontario to provide cheap electricity, as he said, for all its citizens. But primarily the enterprise was a means of attracting industry and providing incentives to develop technologies that utilized electricity, as did, for instance, the chemical industry around Niagara Falls. Ontario Hydro's role in the development of materials and components for nuclear reactors can be seen as a direct descendant of this commitment to industrial growth and infrastructures, financed at public expense.

When the first networks for the generation and distribution of electricity were built in various countries, the political impact of technology on government took a quantum jump. Technologies became political in a new meaning of the word. Now particular technologies became interlocked with particular political goals and trends. The state served technology and technology served the state in a radically new manner.[13] In his book *Networks of Power*, Thomas Hughes demonstrates this by comparing the distribution systems in London, Berlin, and Chicago at the time when electricity began to be a significant factor in public and private life.[14] The early British systems, because of the strong powers of local governments, were composed of a large number of small and locally dispersed generating stations. The German drive toward centralization, which was occurring on the political level, favoured a centralized metropolitan electrical distribution system for Berlin. Although generating stations were privately owned, the very fact that their activities spanned a number of jurisdictions made them part of that drive toward centralization, mainly because of the need to regulate such systems uniformly. One could not, after all, have different voltages, frequencies, or standards between different local jurisdictions.

It's interesting to what extent technical decisions about the operating parameters of these transportation and distribution technologies have political overtones. Some of you will still remember that Ontario Hydro, until about thirty years ago, worked on a 25-cycle standard while the rest of the continent had adopted a 60-cycle frequency for electricity. (One can speculate to what extent Hydro did this to protect its own market and prevent the intrusion of competitors.[15]) There was eventually a very organized, large-scale, and efficient changeover that brought Ontario Hydro into the systems of standards of the North American continent. Standard electrical outlets in North America yield currents of 60 cycles at 110 volts, while Europe and much of Asia work on 50 cycles at 220 volts. Technical parameters like this often constitute basic non-tariff measures designed to protect or expand technologies of a particular origin. In a similar vein there is currently a battle over high-definition standards within the television industry.

One of the reasons I emphasize the link between public policies related to the provision of infrastructures and the spread of technology is the following: Rarely are there public discussions about the merits or problems of adopting a particular technology. For example, Canadians have never been asked (for instance, through a bill before the House of Commons) whether they are prepared to spend their taxes to develop, manufacture, and market nuclear reactors. Yet without publicly funded research and development, industrial support and promotion, and government loans to purchasers, Canadian nuclear technology would not exist. The political systems in most of today's real world of technology are not structured to allow public debate and public input at the point of planning technological enterprises of national scope. And it is public planning that is at issue here. Regardless of who might own railways or transmission lines, radio frequencies or satellites, the public sphere provides the space, the permission, the regulation, and the finances for much of the research. It is the public sphere that grants the "right of way." It seems to be high time that we, as citizens, become concerned about the granting of such technological rights of way.

The design and building of roads and the accesses to harbours and airports have always been paid for by the public. However, underneath the public agenda there is often an agenda that is very specific and sectoral. Public planning for the needs of private industry and for the expansion of technology has gone well beyond the provision of physical infrastructures. There are tax and grant structures, and there is the impact of the needs of technology on the preparation and training of the labour force. Thus, future citizens may gain in computer literacy at the expense of moral literacy or knowledge of history, and it seems to me quite debatable which agenda of education is more in the public interest.

At this point, you may well ask, "What's all this fuss about? Are the infrastructures here depicted as distorting the public sphere — not equally serving citizens and workers? Are they not part of normal life in the real world of technology, needing neither to be questioned nor changed?"

Maybe it is best to cast the answer into a more general framework. For this reason I would like to distinguish between divisible and indivisible benefits and divisible and indivisible costs. This may sound very academic but actually it's not. Let me give you a simple example. If you have a garden and your friends help you to grow a tremendous tomato crop, you can share it out among those who helped. What you have obtained is a divisible benefit and the right to distribute it. Whoever didn't help you, may not get anything. On the other hand, if you work hard to fight pollution and you and your friends succeed in changing the practices of the battery-recycling plant down the street, those who helped you get the benefits, but those who didn't get them too. What you and your friends have obtained are indivisible benefits.

Normally one considers it the obligation of governments, whose institutions are funded through a taxation system, to attend to those aspects of society that provide indivisible benefits — justice and peace, as well as clean air, sanitation, drinkable water, safe roads, equal access to education; public institutions, from courts and schools to regulatory and enforcement systems, developed to do these public tasks. In other words, there is historically the notion that citizens surrendered some of their individual autonomy (and some of their money) to the state for the protection and advancement of the "common good" — that is, indivisible benefits.

Technology has changed this notion about the obligations of a government to its citizens. The public infrastructures that made the development and spread of technology possible have become more and more frequently roads to divisible benefits. Thus the public purse has provided the wherewithal from which the private sector derives the divisible benefits, while at the same time the realm from which the indivisible benefits are derived has deteriorated and often remains unprotected.

Let me put this in concrete terms. The global environmental destruction that the world now has to face could not have happened without the evolution of infrastructures on behalf of technology and its divisible benefits, and corresponding eclipsing of governments' obligation to safeguard and protect the sources of indivisible benefits. Whether the task of reversing global environmental deterioration can be carried out successfully will depend, to a large extent, on understanding and enforcing the role and obligation of governments to safeguard the world's indivisible benefits.

The strength, deep-rootedness, and invisibility of technological infrastructures may offer some explanation as to why the tasks of protecting or restoring the natural environment are so very difficult.

I stressed before that many of the political decisions related to the advancement of technology and the provision of appropriate infrastructures are made on a technical level, far away from public scrutiny. But these decisions do incorporate political biases and political priorities which, in a technical setting, need not be articulated. As far as the public is concerned, the nature of the decision, and its often hidden political agenda, becomes evident only when the plans and designs are executed and in use. Of course, at this point change becomes almost impossible.

One example may suffice here — it concerns the design of the parkways of New York State and the role of Robert Moses, in charge of much of the public works of New York State between 1930 and the late 1960s. His biographer, Robert A. Caro,[16] refers to the bridges and underpasses of the famed New York State parkways. These bridges and underpasses are quite low, intentionally specified by Moses to allow only private cars to pass. All those who travelled by bus because they were poor or black or both were barred from the use and enjoyment of the parkland and its "public amenities" by the technical design of the bridges. Even at the time of Robert Moses, a political statement of the form "We don't want them blacks in our parks" would have been unacceptable in New York State. But a technological expression of the same prejudice appeared to be all right. Of course, to the public the intent of the design became evident only after it was executed — and then the bridges were there.

While I have emphasized here the impact of technology on the nature and task of government and public planning, I do not wish to underemphasize in any way the activities of entirely corporate and sectoral planning. Because of the nature of the infrastructures and the thrust of planning, it is not surprising that public and corporate planning overlap significantly. It is often difficult to see where one ends and the other begins, since there is a communality of expertise. Expert advice comes from professionals who move with ease between governments and technological enterprises. They are educated together. They are usually of the same social class and rarely are they at the receiving end of the plans they devise. In all this citizens find themselves in a difficult situation. Some of the traditional roles of government have changed under the impact of technology, but such changes are rarely articulated in clear political terms. At the same time, political decisions are addressed as technical questions in terms and in places that are inaccessible to ordinary citizens.

In the next lecture I will say more about the problems of being a citizen in today's real world of technology.

IV

In these lectures I have emphasized the difference between holistic and prescriptive technologies because I feel it is important to understand the ways in which prescriptive technologies fragment work. When a task that used to be done by one person is divided into subtasks for a number of people, some basic social parameters change. I said in the first lecture that putting people into a prescriptive mode of work where they have no latitude for judgement and decision-making acculturates them to external control, authority, and conformity. Prescriptive technologies are a seed-bed for a culture of compliance. I have also tried to show how our sense of reality has been changed, especially by the kinds of communications technologies that are based on long-distance information transfer. I introduced the concept of reciprocity to distinguish between the one-way communications so common today and those human interactions based on a give-and-take model.

In the last lecture I showed that since the time of the Industrial Revolution public planning and public resources have provided the infrastructures necessary for the expansion of new technologies and for the diffusion and use of the products of the new industries. This development forged increasingly close linkages between governments and technological growth and development. Our lives today are affected by these linkages. The planning processes which have fostered the development and spread of technology have provided infrastructures that we now consider as a given, normal, and unquestionable part of the real world. Not being mindful of how these structures arose can hamper attempts to change them or to replace old arrangements with new and more appropriate ones.

Today these infrastructures go well beyond road, rail, airport, and power grids; they include financial and tax structures, information networks, and government sponsored research and development on behalf of technological advancement. All these infrastructures could have been designed differently if the first design priority had been human development rather than technological development.[1]

Many technological systems, when examined for context and overall design, are basically anti-people. People are seen as sources of problems while technology is seen as a source of solutions. When, in the factory, the owners feel workers are too slow or too unreliable or too demanding, they are replaced by machines. When students are seen as not sufficiently competent, it is likely to be computers that the school purchases rather than extra teacher's time and extra human help. And when security agencies in this country feel that Canadian citizens harbour thoughts and might contemplate actions that the state doesn't like, they don't invite these citizens to discuss their grievances or alternate thoughts openly and on a basis of equality. Instead, telephones are tapped or files are assembled by purely technological means. And I say that with some feeling because I've been on the receiving end of some of that.[2] The notion that maybe technology constitutes a source of problems and grievances and people might be looked upon as a source of solutions has very rarely entered public policy or even public consciousness.

Now, among all the infrastructures that port specific technologies and their industries, the infrastructures that support the preparations for war and violence are very powerful and deeply entrenched. You note that I do not use the word "defence." This is quite deliberate because if defence, in the sense of maintaining territory and authority, had been the main policy priority, then other structures with other contents would have been developed in Canada and in the rest of the world, at least during the past forty years. There was no shortage of suggestions for these alternative structures of defence, based on civilian defence or what one calls "defensive defence," involving a very different mix of military equipment and quite different international channels of communication.[3] The use of such alternatives, we sadly note, has not happened, and so we need to talk about the infrastructures that support, still today, the preparations for war and for the use of organized violence.

In terms of infrastructures, military procurement is simple and straightforward. The state develops and articulates the needs. The state provides the resources for development and production. The state guarantees final payment and profit, and also provides financial guarantees through loans and research contracts. All this is carried out within a complicated, not to say bizarre, framework of contracts and offsets that exist in order to meet some of the government's non-military needs. Regional development considerations, industrial incentives, and alliances all come into play.

You may say that all this is old hat. After all, didn't Galileo earn a living in Padua by lecturing on the subject of military fortifications? Yes, it is true that the close links between civilian and military powers are ancient, but modern practices have brought new dimensions to these arrangements. Let me just touch on two of them. First, there is what Anatol Rapoport, former chair of the University of Toronto's peace studies program, calls the "technological imperative."[4] In simple terms, it says that whatever can be done by technological means, will be done. The military environment, unconstrained by economic considerations or common sense, offers a particularly tempting field of opportunity for the practitioners of advanced technology. They will provide the newest, most surprising, most novel applications of their expertise, regardless of whether or not these applications address real problems. Remember Star Wars?

My second point is that once a country has embarked on developing an arms production system, it falls upon the government to provide the wherewithal over a long period of time. The development of a weapons system, from design to deployment, may take ten years or more. To keep such technological activities going, public funds have to be committed and expended. To keep the public funds flowing, justifications are needed. And this generates *the need for a credible long-term enemy.*

In the real world of technology, there are then two tasks for the state, if governments wish to use arms production as an infrastructure for the advancement of technology: the state has to guarantee the flow of money, and the state has to guarantee the ongoing, long-term presence of a credible enemy, because only a credible enemy justifies the massive outlay of public funds. The enemy must warrant the development of the most advanced technological devices. The enemy must be cunning, threatening, and just barely beatable by truly ingenious and heroic technologies.

It will be interesting to see how western infrastructures respond to the current situation in the Soviet Union. I would venture that the social and political needs for an enemy are so deeply entrenched in the real world of technology (as we know it today) that new enemies will quickly appear, to assure that the infrastructures can be maintained. I am personally very much afraid that there will be a turning inward of the war machine. After all, the enemy does not have to be the government or citizenry of a foreign state. There is lots of scope — as well as historical precedent — for seeking the enemy within.

Since this lecture was written, the "war on drugs" has broken out. It is not clear to me whether the production of illegal drugs and the trade in them has taken on such a quantitatively different character during the past six months that a declaration of war is in order. However, the bad guys and their helpers have been identified and the enemy's cunning cruelty has been graphically projected onto the local scene. The Red under the bed has been replaced smoothly by the grass in the grass. It is the ease and speed of the transition that I find remarkable.

Finally, the following item, which appeared in the *University of Toronto Staff Bulletin*, supports the point I am trying to make:

National Research Council of Canada has announced a new Canadian program of science and technology. In support of law enforcement, the proposed Canada Police Research Institute will involve the NRC, police, corporate and industrial security, universities, government, and manufacturers of equipment used in police and security work, and will research, develop, and evaluate new security related products. Interested investigators may contact. . . .

In summary, then, this is the situation: most activities in the real world of technology have been planned; the spread of technology has resulted in a web of infrastructures serving primarily the growth and advancement of technology; the presence of these infrastructures and their "forward planning" (often manifested as institutional inertia) severely hinder political or economic changes, even if such changes are viable and appropriate.

For instance, institutional and commercial initiatives to resolve conflict non-violently and to transcend war and violence are hard to sustain in an environment structured by assumptions of escalating violence and warfare, whether economic or military. At one point planning becomes prophecy — not by what planning and structuring does, but by what the plans and structures prevent. The constellation in which "the enemy" has a technological and economic support role profoundly affects the peace movement. It took many of us, who worked for reconciliation and the reduction of hostility, a long time to grasp the power of this constellation.

There is no reason to believe that the support of any state for the preparation for war and violence is the only possible infrastructure to promote a national development of advanced technology. Japan, it should be remembered, was prevented by peace treaties from building up its own armed forces. Japan became — and I would suggest because of, rather than despite, these constraints — a major high-tech success. It developed infrastructures for linking government and industry that are somewhat differently constructed from those of the war-machine builders, but are very effective.

Before leaving the subject, I want to point out the changes that technology has brought to the part of citizens in war preparation and warfare. Just as fewer and fewer unskilled workers are needed in a modern technological production system, a country now has little practical need for raw recruits to operate its modern technological destruction system. Abandoning compulsory military service is not so much a sign of peaceful intentions as it is a sign of galloping automation. But the old pacifist dream that there might be a war and nobody would come and consequently the war could not take place, is no longer valid.

Wars can be started without calling on any additional people. *Military* service from citizens is no longer a prerequisite for war. What is a prerequisite is the compulsory *financial* service of all citizens, well before any military exchange begins. Thus the pacifist's motto "We Won't Fight," must be translated into a new slogan: "We won't pay for the preparations for war and organized violence." This, of course, is the position of Canadians who pay their full income tax but insist on redirecting a portion of it to a peace tax fund so it cannot be used for the war-building purposes I have described.[5]

Before continuing with some thoughts on planning strategies, let me suggest a few general reflections on planning itself. I don't want to talk here about the grand designs of the past — the sort of thing one finds in majestic cities, in palaces and temples; the sorts of layouts that brought a friend of mine to sum up his first impression of Washington D.C. by saying, "The place seems to be designed to be ruins." I want to talk about planning in the modern sense of prescriptive technologies, the kind of thing that Webster defines as "making plans . . . arranging beforehand." I like this simple definition because it says that there are planners as well as *plannees*; there are those who plan and those who conform to what was arranged beforehand. Just as it is easier to give good advice than it is to accept it, it is much more fun to plan than it is to be subjected to plans made by others. The degree of effectiveness of participation by the plannees in long-term planning operations seems to me a true measure of democracy in the real world of technology.

I raise these points because planning is so frequently carried out without the plannees' knowledge or consent. In fact, when plans do not work out, it is often the lack of consideration of the position of the plannees — of their reactions, their counter-planning, their avoidance strategies — that is the reason for the plan's dysfunction. And, as I will stress later, the natural environment is often regarded as a plannee and is usually not consulted.

Even within one plan, there are often contradictory goals. When I was actively involved with the Toronto Planning Board, I wrote a paper called *The Resource Base and the Habitat.* I wanted to point out that large modern cities fulfill two internally contradictory functions. Cities have become the natural habitat for many people; it is in cities that most people grow up, spend essentially all their lives, and bring up their families. Planning is supposed to assure that the city remains a liveable, safe, and sane habitat. But large population concentrations in cities also present a resource base for many enterprises. The need for food and shelter, for entertainment and employment, make cities a resource base like mines or forests. Those who want to exploit the resource base have different planning goals from those who need to develop and maintain the habitat. The resource-base users press for unrestricted access to the resource, and as little responsibility as possible for the debris and residue left by the exploitation of the resource. The garbage heaps of the shopping centres or the plastic containers from the fast-food emporia become the equivalent of a mine's tailing dump and lagoons, left for the community to dispose of.

Different planning perspectives, with their contradictory demands, could be balanced through appropriate democratic planning processes. But although such processes exist in some jurisdictions, their practical execution is frequently weighed against the habitat function. For instance, Sunday shopping is fine for the resource-base constituency, but not so good for the habitat community.

Let me recap again. I have pointed to planning as an activity involving planners and plannees. Planning, in my sense of the word, originated with prescriptive technologies. As prescriptive technologies have taken over most of the activities in the real world of technology, planning has become society's major tool for structuring and restructuring, for stating what is doable and what is not. The effects of lives being planned and controlled are very evident in people's individual reactions to the impingement of planning on them. The real world of technology is full of ingenious individual attempts to sabotage externally imposed plans. As a social phenomenon, such avoidance techniques are well worth studying.

A common denominator of technological planning has

always been the wish to adjust parameters to maximize efficiency and effectiveness. Underlying the plans has been a production model, and production is typically planned to maximize gain. In such a milieu it is easy to forget that not everything is plannable. Actually, most things are properly described by a growth model — and that means many activities of living — and are ultimately not plannable. A quick example from my own experience: Although I was intellectually quite well prepared for the birth of my first child, I was stunned by the degree of randomness that this event created in my life. It took me a while to understand that it was pointless to plan my days the way I used to. This did not mean that I didn't plan or prearrange, but that I needed different schemes to deal with the unplannable.

Women in particular have developed such schemes over the centuries — arrangements that are not a surrender to randomness, but an allotment of time and resources based on situational judgements, quite akin to what I described earlier as the characteristics of holistic technologies. Such schemes require knowledge, experience, discernment, and an overview of a given situation. These schemes are different in kind from those of prescriptive planning. What makes them so different is that holistic strategies are, more often than not, intended to minimize disaster rather than to maximize gain.

Berit Ås, the well-known Norwegian sociologist and feminist, has described this difference in strategies. She sees traditional *planning* as part of the strategy of maximizing gain, and *coping* as central to schemes for minimizing disaster.[6] A crucial distinction here is the place of context. Attempts to minimize disaster require recognition and a profound understanding of context. Context is not considered as stable and invariant; on the contrary, every response induces a counter-response which changes the situation so that the next steps and decisions are taken within an altered context. Traditional planning, on the other hand, assumes a stable context and predictable responses. Planning protocols for prescriptive activities, whether they're industrial, administrative, or educational, can be transferred from one application to another without regard to context.

You may say it's fine to make these academic distinctions between planning strategies, but how would one actually plan to minimize disasters, not in the family, but in the public sphere? I want to give two prominent examples because I am anxious to show that there are indeed no practical obstacles to planning to minimize disaster, and that such approaches are possible in today's real world of technology. One of the examples is drawn from the inquiry led by Thomas Berger into the building of the Mackenzie Valley pipeline;[7] the other is the 1977 study of the Science Council of Canada, entitled *Canada as a Conserver Society*.[8] Again the crux is how context is treated.

All features of the report of the Mackenzie Valley Pipeline Inquiry imply respect for context. The context was illuminated as much as possible, keeping in mind all ramifications of the proposed plan. The circles of consideration were wide, and interactions within the planning milieu were considered important. The recommendations and the proposed course of action made further assessments of the impact mandatory, with built-in revisions to the plan's long-term realization. The Inquiry itself gathered evidence in many different modes, ranging from listening to native residents in their own communities to questioning "experts" on the reliability of forecasts of energy needs or gas reserves. This was a very participatory and interactive process, and it resulted in the recommendation of a ten-year moratorium on pipeline construction, during which urgently needed protective measures for the community and the arctic habitat were to be carried out. The report also recommended certain permanent constraints on future technological activities. Thus the Inquiry resulted in a workable plan to proceed with development while minimizing potential harm. In addition to the recommendations, the Berger Inquiry generated most valuable documentation and research material which would not otherwise have been available in the public realm.

To me the Inquiry is proof that it is indeed possible to engage in a different kind of planning. The complexity of the real world of technology offers no fundamental barriers for implementing strategies to minimize disaster.

The study, *Canada as a Conserver Society*, by the Science Council of Canada, is an example in a slightly different vein. In 1975 a Science Council committee, which I chaired, was asked to explore whether and how Canada could become a conserver society. The council's final report described the concept of a conserver society as follows:

The concept of a conserver society arises from a deep concern for the future, and the realization that decisions taken today, in such areas as energy and resources, may have irreversible and possibly destructive impacts in the medium to long term.

The necessity for a conserver society follows from our perception of the world as a finite host to humanity and from our recognition of increasing global independence.

The realization of the urgent need to minimize disasters became embedded in all phases of the study. This was reflected in the research questions asked and it influenced the conduct of the study itself. From the beginning of the investigation the communities of concern, ranging from citizen groups to regulators and industry, were drawn into the process. During the study, background papers and a newsletter were published and workshops were held. The work of the committee was open and public throughout the period of its deliberations.

The assessment and the specific recommendations incorporated in the council's final report made it clear that planning to minimize disaster is quite feasible within Canada's existing economic system and political infra-structures. It was demonstrated that there was an urgent need to institute public policies related to conservation and to encourage appropriate technological innovations to implement disaster-minimizing opportunities. But the public policy recommendations were not taken up by the government of Canada in 1977. The subsequent onslaught of the gain-maximizing strategies of the 1980s, greatly increased the likelihood of environmental and social disasters and are making the restorative tasks evermore complex and difficult.

The common theme that runs through many disaster-minimizing endeavours is the conviction that ordinary people matter — in the way Schumacher meant when he called his book *Small Is Beautiful; Economics as if People Mattered*. But we must remember that, in the real world of technology, most people live and work under conditions that are not structured for their well-being. The environment in which we live is much more structured for the well-being of technology. It is a manufactured and artificially constructed environment, not what one might call a natural environment. While our surroundings may be a milieu conducive to production, they are much less a milieu conducive to growth.

Speaking about the environment, I must say that I have become more and more reluctant to use this term, although it figures so prominently in current discussions. I feel that "the environment" is now more often a term of befuddlement than a concept that is helpful in the search for clarity. What do people actually mean when they talk about the environment? Is it that constructed, manufactured, built environment that is the day-in-day-out setting of much of the contemporary world of technology? Is it what is euphemistically called our "natural" environment? Why don't we speak about *nature*? It seems such an egocentric and technocentric approach to consider everything in the world with reference to ourselves. Environment essentially means what is around us, with the emphasis on *us*. It's *our* environment, not the environment or the habitat of fish, bird, or tree.

The reluctance to use the word "nature" in political discussions may very well come from a reluctance to acknowledge that there are independent partners on this planet. People are but one part of nature. This recognition is inconsistent with speaking about "our" natural environment, which somehow puts nature into the role of an infrastructure, into the role of something that is there to accommodate us, to facilitate or be part of our lives, subject to our planning. Such a mindset makes nature into a construct rather than seeing nature as a force or entity with its own dynamics.[9]

There are many ways, some of them seemingly small, in which the real world of technology denies the existence and the reality of nature. For instance, there is little sense of season as one walks through a North American or western European supermarket. As a child in Berlin, I still experienced a sense of special occasion when participating in small festive events around the family table to celebrate the first asparagus of the season, the first strawberries, the rare and special gift of an orange in winter. Today such occasions for marking the seasons are rare. Just as there is little sense of season, there is little sense of climate. Everything possible is done to equalize the ambience — to construct an environment that is warm in the winter, cool in the summer — equilibrating temperature and humidity to create an environment that does not reflect nature. Nature is then the outside for "us" who are in an internal cocoon. Indeed, technology does allow us to design nature out of much of our lives. This, however, may be quite stupid. People are part of nature whether they like it or not. Machines and instruments will thrive and work well in even temperatures and constant humidity. People, in fact, may not. For the sake of our own mental and physical health, we may need the rhythm of the seasons and the experience of different climates that can link us to nature and to life.

In no way do I wish to deny the urgent task of "cleaning up the environment," as it is often phrased. But I would like to stress that there is also the urgent task of cleaning up the technocentric and egocentric mindset, to get rid of the notion that nature is just one more infrastructure in the real world of technology.

Sometimes I think if I were granted one wish, it would be that the Canadian government would treat nature the way Canadian governments have always treated the United States of America — with utmost respect and as a great power.

Whenever suggestions for political action are placed before the government of Canada, the first consideration always seems to be "What about the Americans? They may not like it. They may let their displeasure be seen and felt. They may retaliate!" And what about nature? Obviously nature does not take kindly to what is going on in the real world of technology. Nature *is* retaliating, and we'd better understand why and how this is happening. I would therefore suggest to you that, in all processes of planning, nature should be considered as a strong and independent power. Ask, "What will nature do?" before asking, "What will the Americans do?"

Having spoken so much about planning, I do not want to end this lecture without touching briefly on the outcome of such plans. After all, plans are made in order to achieve a particular goal. It is therefore of some interest to see what happens to goals. The retrospective evaluation of plans and predictions is not a particularly comfortable undertaking; maybe this is why it is not well developed as a discipline of study. One would think that much could be learned from finding out why well laid plans and well considered predictions are often so totally out of whack. Of course, the literature is full of anecdotes about how great people so misjudged the development of science and technology that they thought manned flight was impossible, that sound could never be carried by wire, etc., etc. I'm not interested in simply pointing out how far off others have been in their predictions. I have little doubt that some of what I have said here will not stand the test of time either. What I am interested in are reasonable, short-range predictions that did not come true, largely because of a total misreading of context. To me this serves as a reminder that any plausible approach to forecasting the impact of technology must focus on context and involve the experience and views of the plannees as well as the planners who design the technology.

In 1964 the British journal *The New Scientist*, under its editor, Nigel Calder, approached social and physical scientists from a large number of countries and requested from them a prediction of what their fields or their countries would be like in 1984, just twenty years in the future.

These were scholars who knew their fields well and who were looking ahead to the Orwellian date of 1984. The responses were published in 1964 and the book is still in print.[10] I warmly recommend it to you because it's quite stunning. There are visions of air-conditioned plenty, of creative leisure and broadly based political involvements. But I especially like the prediction of a senior official at IBM, in an article called "The Banishment of Paperwork." He confidently forecast the total absence of paperwork in 1984: Computers, within two decades, would have become the sole medium of communication, while all that burdensome paper would have vanished from our desks.

In the case of Canada, the then vice-president of the National Research Council entitled his contribution "Canada: Plenty of Room for People." He predicted a Canada of at least thirty-five million people, exporting wheat, pulp and paper, iron ore, nickel, and many other metals. At the same time Canada's manufacturing industries would be thriving. His prediction also foresaw a great future for educational television. "For instance, by dialling the public library, one might be able in 1984 to read any book while sitting in one's house, the printed page presented on the television screen. The blind, the lazy and the illiterate can listen instead." He further anticipated that the ratio of motor vehicles to people would probably stabilize before 1984, for the simple reason that cars would be so numerous that the increment of usefulness for additional vehicles would be essentially zero. Every place would be so clogged up that more cars would be pointless. The lack of roads on which to drive cars and parking lots and garages in which to leave them would make more cars virtually unwanted.

What is so striking in this and many similar comments is the lack of appreciation of the political dynamics of technology. Of course roads will be built even if they gobble up the best agricultural land. Of course garages will be erected in preference to housing for the poor.

The collection contains other quite astonishing predictions — astonishing in their belief that the growth of technology and information would bring limitless comfort and potential prosperity. The title of one essay, "Bioengineering — Opportunities without Limit," is representative of such an outlook.

There are many contributions foreseeing air-conditioned and totally climate-controlled living and working spaces — and one senses the wish to keep something out, to protect and encapsulate people. But the contributions contain little awareness of the connectedness of life and the fact that whatever may harm humans will harm air, soil and water, plants and animals.

There are acknowledgements of hunger in the Third World — a number of the contributors were from outside Europe and North America — and of the need to nourish a growing world population. But there is no recognition whatever of the potential for hunger and poverty in the First World. The reality of *economic* underdevelopment was perceived by the scholars; the reality of *moral* underdevelopment was rarely mentioned.

When I talked earlier about the need to look at technology in context, I meant the context of nature and people.

When predictions turn out to be as wrong as many of those in *The World in 1984*, it is because context is not a passive medium but a dynamic counterpart. The responses of people, individually and collectively, and the responses of nature are often underrated in the formulation of plans and predictions. Electrical engineers speak about inductive coupling: A changing field induces a current, which may induce a counter-current. Change produces changes, often in different dimensions and magnitudes. Maybe what the real world of technology needs more than anything else are citizens with a sense of humility — the humility of Kepler or Newton, who studied the universe but knew that they were not asked to run it.

V

In each of these lectures I have stressed that the division of labour characteristic of prescriptive technologies has resulted in the acculturation of people into a culture of conformity and compliance. And that has many very significant consequences.

Ivan Illich pointed out in his 1981 essay, *Shadow Work*,[1] that prescriptive technologies, particularly those in the administrative and social-service sectors, produce the desired results only when clients — for instance, parents, students, or patients — comply faithfully and to the letter with the prescriptions of the system. Thus, advanced applications of prescriptive technologies require compliance not only from workers, but also from those who use the technologies or are being processed by them. Illich stressed the role of individual and group compliance by citizens in this process of making prescriptive technologies work. In my last lecture on the infrastructures that support technology, I pointed out that not only do individuals have to fit into the compliance scheme, but so do governments and public institutions.

All social interactions proceed according to a certain characteristic internal logic. Actions are carried out in a particular manner in the expectation of a commensurate reaction, whether it is the tit-for-tat logic of strife or a turning of the other cheek. Where prescriptive technologies are structured to perform social transactions, these transactions will be organized or reorganized according to the logic of technology, the logic of production. Thus, as more and more of daily life in the real world of technology is conducted via prescriptive technologies, the logic of technology begins to overpower and displace other types of social logic, such as the logic of compassion or the logic of obligation, the logic of ecological survival or the logic of linkages into nature. Herbert Marcuse, in *One Dimensional Man*, speaks of this overpowering.[2]

In order to clarify the mechanisms by which the logic of technology suffocates other forms of social logic, I want to look into the patterns that emerge during the introduction of different technologies into a society. Historians of technology point out that there are general stages in this process of introduction. Inventions and innovations may lead to the development of a particular technology; this, in turn, can bring growth of the technology, social acceptance, and standardization of production as well as products. Standardization usually results in technical as well as economic consolidation. At this point, a technology can become so stabilized that there is no room for further technical inventiveness. The technology itself is no longer changing; at best it will be marginally modified.

One can also look at that same process of invention, growth, acceptance, standardization, and stagnation from the point of view of the society that accepts the technology. This view yields a much richer picture. From this vantage point we see that in the early stage of a particular invention, a good deal of enthusiasm and imagination is generated. There are efforts to explore and explain just how wonderful and helpful the new invention will be. Science fiction often gives a framework for such imaginary explorations. The dreams of flight, of fast private transportation, of instant communication across continents, and of helpful machines, all stress liberation from hard physical labour at work or drudgery at home. Wellsprings of creativity and freedom from toil seem to be just around the corner. In this phase technologies create human bonds and a sense of excitement in people who feel grateful to be part of such wonderful, progressive times. The voices of reservation sound like disgruntled skeptics, fearful of change — like the old lady who said that if God had wanted us to fly, she would never have given us the railways.

During the Industrial Revolution there were "Odes to the Steam Engine"; the great exhibitions at the turn of the century were filled with public displays of light and sound and they depicted a much better and more exciting life almost within the reach of ordinary people.[3] In the course of this phase of youthful exuberance, technologies achieve broadly based entry into the public mind and the public imagination. Yes indeed, people feel, it might be nice to try out a car or a telephone, or work with some of these fancy machines.

After this phase, with its flights of imagination, human contacts, and excessive hopes, a new phase appears. This is the phase of the stern father saying, "What do you really want to do when you grow up?" This is the phase of growth and standardization of the technology. From here on the involvement of people, whether workers or users, is drastically reduced.

Take, for instance, the motor car. In its young phase, it could be quite appropriately called a "mechanical bride," the term used by Marshall McLuhan to describe the relationship between car and owner.[4] Care was regularly lavished by young men upon their vehicles, polishing and tuning them, repairing them and improving their performances. There was a sense of camaraderie among the owners and they would admire each other's mechanical brides. Little is left of this era in today's real world of technology. In the automobile's technological middle age, it is hard, if not impossible, to tune or repair one's own vehicle — other than by taking out interchangeable parts and popping in identical replacements. Cars are costly to buy and to maintain, but for many owning a car is neither a joy nor a luxury but a necessity. The mechanical bride has turned into a demanding but essential business partner.

Technical standardization of cars has occurred, and with it the elimination of the user's access to the machine itself. At the same time, the infrastructures that once served those who did not use automobiles atrophied and vanished. Some may say they were deliberately starved out. Railways gave way to more and more roadways. And thus a technology that had been perceived to liberate its users began to enslave them. The real joy of owning wheels, the sense of independence that allowed drivers to go wherever and whenever they wanted to go, became muted because, in reality, there were usually thousands and thousands of others who wanted or had to go at the same time to the same place.

The early phase of technology often occurs in a take-it-or-leave-it atmosphere. Users are involved and have a feeling of control that gives them the impression that they are entirely free to accept or reject a particular technology and its products. But when a technology, together with the supporting infrastructures, becomes institutionalized, users often become captive supporters of both the technology and the infrastructures. (At this point, the technology itself may stagnate, improvements may become cosmetic or marginal, and competition becomes ritualized.) In the case of the automobile, the railways are gone — the choice of taking the car or leaving it at home no longer exists.

When assessing the individual and social impacts of different technologies, the internal chronologies outlined above must be considered. For instance, within my memory not only cars but audio equipment had a youthful and involved phase, when people talked about matching impedances, building turntables and preamplifiers from kits, and comparing their respective successes in achieving high-fidelity sound reproduction. That phase is gone, too. The market now offers standard plug-in components and owners may listen — but not intervene beyond pushing buttons. We're in the midst of a similar evolution in the use of computers. Today's popular and emotional involvement with the personal computer is very much akin to the mechanical bride phase of the automobile. Once again there is a promise of liberation — there's no need any more for good typing skills or correct spelling, for arithmetic or knowing how to do percentages, no need to keep the files straight. The computer can do it all. It can even recycle that painful letter to Aunt Amelia to other members of the family with only slight modifications.

Manufacturers and promoters always stress the liberating attributes of a new technology, regardless of the specific technology in question. There are attempts to allay fear, to be user-friendly, and to let the users derive pride from their new skills. There is an effort to build up user communities brimming over with warm feelings of sharing newly won expertise. In the computing field the best barometer for these endeavours is the large number of popular computing magazines. They range from the free-of-charge ones, such as *Toronto Computes*, that can be picked up in the neighbourhood convenience store, to the glossy numbers like *PC World*. All of them have similar styles based on a gushy, breathless, gee-whiz kind of journalism. They remind me of nothing so much as the

eatshops that exploited the labour of women
arly the labour of women immigrants. Sewing
came, in fact, synonymous not with liberation
ploitation. The sewing machines at home were
s machine-sewn household goods and garments
e readily available on the mass market. These
were produced by the prescriptive technologies
ed a situation in which one seamstress only
sleeves, another worker put them in, another
ttonholes, another pressed the shirts. A strictly
ive technology with the classic division of labour
m the introduction of new, supposedly liberating
tic" machines. In the subsequent evolution of the
t industry, much of the designing, cutting, and
ling began to be automated, often to the complete
on of workers.

social history of the industrialization of clothing
ilar to the current phase in the industrialization of
. Food outlets put frozen or chemically prepared
meals" together like sleeves and collars for shirts —
are "Mc-Jobs" and no security of employment.
ed, women sew less, cook less, and have to work hard
ide the home to be able to buy clothing and food.

What turns the promised liberation into enslavement
not the products of technology *per se* — the car, the
mputer, or the sewing machine — but the structures
d infrastructures that are put in place to facilitate the use
these products and to develop dependency on them.
he funny thing is that in the course of this process of
preading technology, the ordinary things — a home-

women's magazines that one used to pick up in great
numbers at supermarkets, when fancy kitchen appliances
and prepared foods were being introduced. Here too were
coupons and free gifts, recipe exchanges and user-proven
tips for shortcuts or special effects. I urge you to look at
some of the computer magazines from the point of view of
set-up and style. You'll find columns on "How I found
ten more uses for Lotus 1-2-3" and "How I extended the
capacity of my Mac beyond imagination." These columns
strongly remind me of columns like "How I used a packaged cake mix and even my mother-in-law didn't notice it."

You may well question the validity of this analogy and
ask, "Is there really a correspondence between the ways in
which prepared and packaged food was introduced and
the marketing of home computers?" I think there is,
and the argument goes as follows: The post-war period
saw considerable advances in the chemistry of food
additives, allowing for much longer shelf-lives of products
to which these chemicals had been added. At the same
time, new machinery made the individual packaging of
goods economically viable. New distribution systems —
particularly by air — opened up around the same time.
Thus foods could be chemically stabilized, industrially
packaged, and commercially shipped over long distances.
The challenge to advertising and marketing was to
stimulate and entice changes in cooking and eating habits
that would utilize the new products (and the appliances
to go with them).

The chain of evidence then goes from "Betty Crocker"
and her recipes, to magazines and gadgets, to frozen food

and TV dinners, to today's attempts to market irradiated food (or infant formula in the Third World). The promotion of industrially processed food was geared to make women accept the new products as a liberating and exciting addition to their lives without worrying about chemical additives or increasing costs. I see a very similar scenario in the promotion of computers for individual use, and I think the pitch is quite deliberate. It is aimed at creating an atmosphere of harmless domesticity around the new technology to ease its acceptance. Who could fear these cute and clever things that make life so interesting at home when the kids play games on them?[5] There's also the language of computers to support this image of harmless domesticity. One speaks of booting up and boilerplates; one talks about mouse and menu. The user has the feeling of choice and control, of mastery and a comfortable relationship with the machine and with other users.

But this phase will not last. Behind that pink fluff one already sees the features of global restructuring. The changes in the workplace are there and it is not the workers who exercise control. After you've looked at the gushy computer magazines, you may want to read Heather Menzies' book, called *Fast Forward and Out of Control*,[6] in which she speaks about global restructuring in terms of the Canadian economy and Canadian workers.

If one doesn't watch the introduction of new technologies and particularly watch the infrastructures that emerge, promises of liberation through technology can become a ticket to enslavement. I'd like to remind you of

one example of the d
new technology. The c
Let's look at the introdu

In 1851 the mechar
commercially available d
a household appliance th
chores and drudgery of l
sewed at home for their fa
working for others, the pro
Not only were individual wd
device, but there were high ho
The following paragraph, wr
Cheris Kramarae in an article o
machine:

> The sewing machine will, after s
> banish ragged and unclad human
> all benevolent institutions, these m
> operation and do or may do 100 tin
> clothing the indigent and feeble tha
> of all the charitable and willing ladie
> the civilized world could possibly pe

The authors of this prognostication ev
that the introduction of the sewing mach
in more sewing — and easier sewing — by
always sewn. They would do the work th
done in an unchanged setting.

Reality turned out to be quite different.
of the new machines, sewing came to be don

cooked meal, an individually made garment — become prized and special, while what had been prized and extraordinary — for instance, cloth or fruit from the Orient — appears now to be quite ordinary and routine.

To recap: many new technologies and their products have entered the public sphere in a cloud of hope, imagination, and anticipation. In many cases these hopes were to begin with fictional, rather than real; even in the best of circumstances they were vastly exaggerated. Discussion focused largely on individuals, whether users or workers, and promised an easier life with liberation from toil and drudgery. Discourse never seemed to focus on the effects of the use of the same device by a large number of people, nor was there any focus on the organizational and industrial implications of the new technologies, other than in the vaguest of terms.

In spite of the exaggerated individual promises, techniques were treated as if they would fit easily into "normal life." Carefully selected phrases used to describe new technical advances could generate an image of chummy communities and adventurous users. But once a given technology is widely accepted and standardized, the relationship between the products of the technology and the users changes. Users have less scope, they matter less, and their needs are no longer the main concern of the designers. There is, then, a discernable pattern in the social and political growth of a technology that does not depend on the particular technical features of the system in question.

It should be evident by now that there is no such thing

as "just introducing" a new gadget to do one particular task. It is foolish to assume that everything else in such a situation will remain the same; all things change when one thing changes. Even the introduction of a dishwasher into a family's life changes their communication and time patterns, their expectations and the ways in which the family works together.

Feminist discourse and research has shed much new light on the social and political dimensions of science and technology. Inquiries ranging from *Reflections on Gender and Science* by Evelyn Fox Keller to Cynthia Cockburn's *Machinery of Dominance* have brought fresh perspectives and new knowledge to questions about the social structuring of science and technology. The resulting insights have led to greater clarity about the nature of the enterprise we call science and technology. There exists now a substantial body of documentation showing how teaching, research, and practice in most areas of science and technology follow essentially male patterns by being basically hierarchical, authoritarian, competitive, and exclusive.[8] In terms of technology such findings should not come as a surprise. Major facets of technology are related to prescriptive practices and thus to the development of instruments of power and control.

We have seen in the example of the parkway systems of Robert Moses the technological packaging of racial and class bias. There is also a great deal of technological packaging of gender bias. In the first of these lectures, I pointed out how technology as practice can define and identify content and practitioners. The perceived gendering of tools and professions leads to strong biases against women, biases that may be personal as well as institutional. Scholars, for instance Sally Hacker and Margaret Benston, have documented such biases and their consequences.[9]

One must not think about these biases solely in terms of being hurtful to women, however. The exclusion of women as formative practitioners of technology is even more harmful to society. One of the most powerful barriers to the creative participation of women in technological activities is the fragmentation of technical work and its rigid structuring. In 1984 I wrote a paper entitled "Will women change technology or will technology change women?"[10] In it I addressed the question of the technological structuring of knowledge and work. As I've done in these lectures, I outlined in the paper the nature of prescriptive technologies and interpreted what working within them entails. I contrasted this mode of working to women's historical experience of situational and holistic work. The success of such work depends strongly on personal judgement, on knowledge of the total work process, and on the ability to discern what the essential variables are at any one time. None of these attributes of knowledge and judgement are required in modern industrial production and, in fact, they're usually not appreciated in workers. Nevertheless, these are the skills and the strengths which women very often bring to the workplace.

I argued in the paper that it seems pointless to bring women into the real world of technology merely to work in the existing technological mode. The great contribution of women to technology lies precisely in their potential to change the technostructures by understanding, critiquing, and changing the very parameters that have kept women *away* from technology. Only then do we have the possibility of changing the real world of technology itself. Happily, I am seeing small beginnings of such structural changes — work that is less prescriptive, workplaces that are less hierarchical, relations that are less rigidly ranked. However, it's barely a start. In the real world of technology, women are still most likely to succeed if they become "one of the boys" just as fast as they can.

Much as I would like to say more about the dimensions that science and technology have lost because of their hierarchical structures and because of the absence of women, I now need to speak of women as workers during the introductory phases of new technologies. Standard histories of technology[11] rarely acknowledge the contributions of women to the development and spread of modern technologies. Yet it is entirely fair to say that without the work of women, their willingness to do extremely delicate but repetitive jobs, and their ability to learn intricate work patterns, the electrical and electronic technologies could not have developed in the way they did. It has often been stressed how poorly women were paid in the new technological order; it has been stressed much less how essential their skill and perseverance have been for the development of the technologies themselves. When one reads about the history of the manufacture of electrical and electronic equipment, it is quite clear how crucial the presence of women workers was to the manufacturers' success.

The same observation holds true for the transfer of mechanical technologies to the office. Who but women would take up those monstrous early typewriters and learn to type faster than people could speak?

A politically most interesting case of women in the new technological world is the history of telephone operators. In the early days of the telephone, operators had much larger tasks than just connecting wires through a switch-board. And telephone operators, in contrast to telephone engineers or repairmen, have always been female.

In 1988 Carolyn Marvin published a most insightful and original book depicting the reactions of both the general population and the business community to the introduction of electrical technologies ranging from electric lights to the telegraph and telephone.[12] *When Old Technologies Were New* documents the interplay between the introduction of the new technologies and changes in social relations. The book also illustrates the phase of wild imagination, of exaggerated hopes for and irrational fears about the new technical developments that I mentioned earlier in this lecture. Marvin provides ample evidence of the intense involvement of individuals in using, and through their use in modifying, the new electrical technologies. This involvement took many forms but I would like to focus here on the role of the operator in the use and development of telephone systems.

In order to establish the technology, it was necessary to find and develop appropriate uses for the telephone and to establish these uses as part of a normal way of living and doing business. The telephone operator was the link between the new technology and the community. Let me stress that the operators were not mechanical or electrical links; they were human links. Telephone operators provided the operational centre that directed the technology as much as it facilitated its applications. The operators found ways to make the technology useful, and they were filling the role of those who today would be called "product-development engineers."

In the early stages of development, many telephone switchboards were also telegraph stations. In 1892, for instance, the results of the U.S. presidential election were telegraphed to switchboards and from there operators gave, every hour on the hour, election results to their telephone subscribers. By the same token, telephone subscribers could phone in for recent sports results. Telephone operators could take and relay messages and they could link up with other operators. In fact, you could do a conference call in 1890. Telephones at the turn of the century provided more than two-way individual communication. A reporter on a sports field could describe an important match and the phone brought back to him the cheering and booing of those who listened to the phone on this giant party line, connected just for the occasion. The listening audience was large, in the range of many thousands. Some events were, in fact, "broadcast" over the telephone. In Paris one could, for five *centimes*, listen to half an hour of The Paris Opera, and follow the performance by phone "as it happened."

During this phase, in which various applications of telephone and telegraph communication were developed and tested, the operators were the central participants in the experiments. One could at that time not imagine the telephone working without an operator. The operator's role was that of an operating and trouble-shooting engineer as well as that of a facilitator.

Once the development and the social integration of the technology had been accomplished to the satisfaction of its promoters, once the infrastructures of needs had been established and alternatives had been eliminated, the technology began to remove the human links. Switchboards were redesigned and automated, operators were slowly supplanted by sophisticated equipment; and together with the human links, the community uses of the telephone began to disappear. Today engineers are busy inventing and selling devices that fill the functions of the operators. We have answering machines and scheduling devices on our phones. We try, often unsuccessfully, to electronically transfer calls within our own organizations; conference calls can be arranged.

A lot of what happens on the phone now is one-way communication. We may be able to receive a telephone weather report, but we can't say to the recorded announcement, "Are you sure? It looks like rain to me." On phone-in radio programs, there is at best one-to-one communication. One can neither participate in the discussions of others nor cheer when somebody else makes a good point during the program. There are times, of course, when one-to-one communication on the phone is essential and lifesaving. The distress lines that are kept busy in many big cities are one example. But telephone technology also brought us a kind of ultimate parody of electronic non-communication. Here's how it's advertised — on television, of course:

> "The two best places in town to meet fun and exciting people — for friendship, romance, and fun. It's the Dating Game. Listen to personal ads, send and receive messages instantly over your touch-tone phone. Call 1-976-9595. In the mood for a party? Want to meet new friends or perhaps that special someone? Call the Party Line, 1-976-8585. The Dating Game. The Party Line. Three dollars per call, plus toll if any."

These lines are apparently very popular with young people — affluent young people, I would say. I am told that many of them just listen in, and even on the Party Line, where they could talk to others, choose not to do so. This 976 line is part of the real world of technology. What does it say about our society, when human needs for fellowship and warmth are met by devices that provide illusions to the users and profits to the suppliers?

The reason I find this particular application of technology so upsetting is that as a response to loneliness, it seems to me deceitful and fraudulent. There are no shortcuts to the investment of time and care in friendship and human bonding, and it is fraudulent to pretend otherwise. When human loneliness becomes a source of income for others through devices, we'd better stop and think a bit about the place of human needs in the real world of technology. What happened to the fuller and more creative life we were supposed to get when technology finally allowed the "human use of human beings"[13] (as Norbert Wiener put it in 1950).

Just to underline again how frequently technological development in the real world discounts the human dimension, I want to mention another group of women who, like the early telephone operators, had a central role in developing and testing a new technology without getting much credit or reward from it. These are the office workers and secretaries who began to establish the mechanical office, the predecessor of today's electronic office. They too were product-development engineers who found ways to use such unruly creations as the early duplicators and adding machines, the newly developed typewriters and tabulators. Theirs was often a tough turf, as Elaine Bernard illustrates in a paper on the history of the typewriter.[14]

The typewriter that E. Remington and Sons developed when the company was diversifying due to a declining market for precision rifles was a piece of machinery that was pretty unsuited to the task at hand. Early typewriters experienced difficulties with jamming hammers and keys; this happened especially when the letters that were typed were located close to each other on the keyboard and on the bank of hammers that the typist activated. As the operators began to type faster, the keys couldn't rise and fall back quickly enough and got tangled.

So Remington commissioned a study into the frequency of letter associations. From this the designers would know which keys were likely to be used sequentially. On the basis of such information, a new keyboard and a new arrangement of hammers was designed. Now the keys and hammers that would be used in sequence during typing were physically separated as widely as possible. This meant much more effort had to go into the typing, but the problem of jamming was "solved" as a technical issue. Thus mechanical design considerations, rather than the ease of typing, have given the world the peculiar keyboard that is even today on all typewriters and terminals. The problem of jamming keys has long since disappeared, but generations of typists and keyboard users have been stuck with the Remington layout although substantially more convenient keyboards have long existed.

These episodes from the history of the telephone and the typewriter say something about the place of the operators in technical development and their essential role in making technology work. But they also tell us some- thing about the disregard that technical designers can have for the needs of operators. Typists not only got awkward machines, but they — and the telephone operators — also encountered the usual division of work that has become part of mechanization and automation. As the technologies matured and took command, women were left with fragmented and increasingly meaning-less work.

In closing let me read a poem by Helen Potrebenko that illuminates this situation:

ANOTHER SILLY TYPING ERROR

The nature of typing is such that
there are none but silly errors to make:
renowned only for pettiness
and an appearance of stupidity.
I don't want to make silly little errors;
I want to make big important errors.
I want to make at least one error
which fills my supervisor with such horror
she blanches and almost faints
and then runs to the manager's office.
The manager turns pale and stares out the window
then resolutely picks up the phone
to page the big boss at his golf game.
Then the big boss comes running into the office
and the manager closes his door
and hours go by.
The other women don't talk
or talk only in whispers,
pale as ghosts but relieved it isn't them.
An emergency stockholders' meeting has to be
    called
about which we only hear rumours.
To make sure I don't accidentally get a job
with a subsidiary, allied company, or supplier,
I am offered a choice of either

fourteen years severance pay or early retirement.
A question is asked in Parliament
to which the Prime Minister replies by assuring the
    House
most typists only make silly typing errors
which only rarely affect the balance of trade.
The only time I get to talk about it
is when I am interviewed (anonymously) for an
    article
about the effect of typing errors on the economy.[15]

VI

It is hard to imagine one's own time as history. Or to think that someone will examine the artifacts of our own time with as much pleasure as I experience when I examine ancient objects. Yet it will happen, and our artifacts will reflect our values and choices, as artifacts have done throughout the ages.

Let me cite a historical example because I want to stress again that no technology is God-given. The notion that the technical requirements for an efficient operation dictate the way technology is laid out is usually not correct. The way a task at hand is dealt with can change as the values and priorities of a society change. My example involves slag from an ancient Peruvian smelting site. A colleague of mine had excavated a large site. It yielded all sorts of interesting smelting-related technical artifacts, ore and slag, tools and furnace parts. My friend visited me to show me the slags from the site. When we looked at them under the microscope I realized that they were different from any other copper-smelting slags I had seen before — ancient Roman or Chinese, European or Middle Eastern. (One would think that when it comes to copper slags, if you have seen one, you have seen them all.) First I did not believe that what we were looking at was slag at all — and I remember saying, "This isn't slag, this is a dog's breakfast."

Eventually the residue made sense, as evidence of the way such processes were carried out in other societies. In ancient Peru most ordinary people were subject to a labour tax — that is, they gave time to agricultural or other projects of the state or the local community.[1] These people were not specialists or particularly skilled at the task at hand. The copper smelting, as the slags showed, was appropriately laid out for these conditions. There were many quite small furnaces, not larger than soup pots, which could be filled with an appropriate mixture of ore and charcoal and heated, using mainly unskilled labour. The little furnaces were left to cool, then broken up — with some of the contents indeed looking like a dog's breakfast. The mixture of slag and copper could be further broken up using water and panning techniques. This type of smelting could be carried out effectively with non-specialized labour. Only when it came to the remelting and alloying of the recovered copper would skilled artisans — "the experts" — be needed.

A process laid out like this would not occur to us — nor

would it have occurred to the Romans — but it made good sense in the context of ancient Peru.

To be sure, history is not a rerun for slow learners. We are not ancient Peruvians. But I cite this historical example to help expand our discourse and our social imagination. Technology is not preordained. There are choices to be made and I, for one, see no reason why our technologies could not be more participatory and less expert-driven.

In this lecture, then, I would like to concentrate on reflections about change. First, I would like to sum up the flow of my exploration of the real world of technology by reviewing some of the concepts I have introduced. I would like to discuss what I think needs to happen if the real world of technology is to become a globally liveable habitat. And finally, I want to suggest some practical steps in this direction that are most urgent and doable. But don't expect from me a blueprint to cure the faults of all other blueprints. From what I have said during these talks it will be clear that it is the nature of blueprints, of prescriptions and plans, that is at issue here, not the details of one or another scheme.

Time will not permit me to speak about the work of Ilya Prigogine and his associates,[2] or to introduce you to the thinking of C. S. Holling on the responses to changes of living systems and on ecosystems management.[3] But I see in the contributions of these scholars some of the new and helpful non-blueprint concepts that might allow us to proceed with constructive alterations to the house that technology has built and in which we all live.[4]

I began this discourse with technology as practice, which allowed me to link technology to culture — culture defined as commonly shared values and practices. The way of doing something can be "holistic" when the doer is in control of the work process. The way of doing something can also be "prescriptive," when the work — whatever it might be — is divided into specified steps, each carried out by separate individuals. This form of division of labour, historically quite old and not dependent on the use of machines, is a crucial social invention at first practised in the workplace.

Authors who have discussed the rise of bureaucracy have neglected to examine the structures of the workplace.[5] It is the acculturation into a culture of compliance built on the willing adherence to prescription and the acceptance as normal of external control and management that make bureaucracy possible. An understanding of the nature of prescriptive technology and the social consequences of the division of labour is important to the appreciation of the speed and strength of the spread of technology. In earlier lectures I also introduced the concepts of a growth model and a production model as schemes that underlie discourse and decision-making in matters of technology. We talked about the separation of knowledge from experience that science has brought. In its wake came the rise of experts and the decline of people's trust in their own direct experiences.

We looked into the so-called communications technologies and how they drastically altered the perceptions of reality. Within a very short historical period these technologies have affected perceptions of space and time and have led to new pseudorealities and pseudo-communities. I stressed the concept of reciprocity and pointed out that modern technologies are frequently designed to make reciprocity impossible. In such situations human responses can neither be given nor received. The absence of reciprocity turns many communications technologies into non-communications technologies.

The role of governments in the promotion and support of technology changed drastically after the Industrial Revolution. Since then publicly financed infrastructures ranging from railroads to electrical distribution networks and financial and tax structures have emerged. They are largely support systems for the advancement of technology; without them, the development and acceptance of inventions such as the telephone, the automobile, and the computer could not have taken place.

At this juncture I stressed the distinction between divisible and indivisible benefits and costs and pointed out how the ongoing provision of technological support structures has been accompanied by a neglect by governments of their traditional mandate to safeguard "the commons" as a source of indivisible benefits. I have discussed how often infrastructures that are publicly funded have become roads to divisible benefits, venues of private and corporate profit. At the same time, those things we hold in common (as sources of indivisible human benefits), such as clean air and uncontaminated water and natural resources, are less and less safeguarded by those who have been given authority to govern.

With the predominance of prescriptive technologies in today's world — technologies that have taken over like a giant occupation force — planning has become the major policy tool. Basically, it doesn't matter whether one considers governmental or corporate planning. The difference is not always easy to ascertain. What is important here is to realize that there are planners and plannees, that is, there are those who develop plans and those who have to conform to them.

It is equally important to realize that there are, in principle, two different planning strategies. There is planning in order to maximize gain, and there is planning in order to minimize disaster. I gave two examples of public planning for the latter purpose — the Mackenzie Valley Pipeline Inquiry[6] and the report by the Science Council of Canada, *Canada as a Conserver Society*[7] — in order to show that planning to minimize disaster is possible, not only in theory but in practice.

In this context I felt it was necessary to stress that the concept of "the environment" includes two separate components. One refers to the built and constructed environment, which is truly a product of technology; the other is *nature*, which is not. I made a plea that we get away from the egocentric and technocentric mindset that regards nature as an infrastructure to be adjusted and used like all other infrastructures. I said that if I had one wish, I would wish that the government of Canada would treat nature with the same respect with which all governments of Canada have always treated the United States as a great power, and a force to fear. When suggestions for political actions are given to the Canadian government the first response is often "What will the Americans say?" I really wish we would look at nature as an independent power and when planning ask, "What would nature say?"

Finally, I tried to show some common patterns that occur when new technologies are introduced into society. We looked at cars and the three ages of response to them, and at computers, industrialized food, and sewing machines. I pointed out how often the promise of liberation in the first stages of the introduction of a technology is not subsequently fulfilled, and that there is quite a sophisticated mechanism of building up dependency after having built acceptance of the new technology.

In light of all this, it would be an understatement to say that all is not well in the real world of technology. The social, economic, and human costs of technological advances are in evidence all round the globe. They are described in newspapers and magazines, discussed on radio, television, and in a stream of books. This ceaseless exposition of problems surrounding modern technology has gone on for at least the last two decades. I've looked again at Fritz Schumacher's "Technology for a democratic society," a talk he gave at a conference in Switzerland in September 1977, the day before he died suddenly of a heart attack.[8] Schumacher's talk contained much, if not all, one needs to know in order to proceed with the task of rectifying the misuses and inhumanities of modern technology. He stressed context; after all, his notions about the appropriateness of technology are based on the recognition of the centrality of context.[9] In the speech,

Schumacher used wordplay on his name, which means "shoemaker." He reminded his audience that a good shoemaker not only needs to know about making shoes, but also has to think about *feet* and how they may be different, because in the end the shoe has to fit the foot. The main theme of Schumacher's last talk, though, was his concern about technology as "a force that forms society and today forms it so that fewer and fewer people can be real people."

It seems, therefore, fair to say that the convincing and urgent case for not proceeding with global technological expansion along the then established patterns was made at least twenty years ago (and continues to be made with stronger and stronger arguments). Nevertheless, there has been no change in direction over the last twenty years, but rather an acceleration of technological development along the lines known to lead to greater and more irreversible problems.

We then have to ask, "Why did nothing substantial happen?" Or we can turn the question around and ask, "What will it take to initiate genuine change?" I would like to suggest to you that the crisis of technology is actually a crisis of governance. I say governance rather than government because I think the crisis is much deeper than the policies of any particular government, although some governments are worse than others. The real crisis, I think, can best be addressed if you ask yourself, "What is the task of the government in this real world of technology? What are the tasks for which we elect and pay governments? What do we expect them to do, rather than to say?" And it would seem foolish to assume that in a world in which technology has changed all practices and relationships, the practice of government and the relationship between the governed and those who govern would remain unaffected. Kids in school are taught that democracy means government by the people, for the people, but the major decisions that affect our lives, here and now in Canada, are not made by the House of Commons or as result of public deliberations by elected officials. I hold that, in fact, we have lost the *institution* of government in terms of responsibility and accountability to the people. We now have nothing but a bunch of managers, who run the country to make it safe for technology.

What is needed, then, is to change or reform the institution of government in terms of responsibility and accountability to the people as people. And how could this be done? It cannot be done by some great leader or guru coming out of the woodwork somewhere, and I am most anxious that none of what I say should be interpreted as an invitation to Fascism. What needs to be done cannot be done as a dictate from on high but will come as an inescapable consequence of movements from below.

I have long subscribed to what I call Franklin's earthworm theory of social change. Social change will not come to us like an avalanche down the mountain. Social change will come through seeds growing in well prepared soil — and it is we, like the earthworms, who prepare the soil. We also seed thoughts and knowledge and concern. We realize there are no guarantees as to what will come up. Yet we do know that without the seeds and the prepared soil nothing will grow at all. I am convinced that we are indeed already in a period in which this movement from below is becoming more and more articulate, but what is needed is a lot more earthworming.

Now, how do we do that? First of all, it is necessary to transcend the barriers that technology puts up against reciprocity and human contact. One of the reasons why I dwelt so much on non-communications technologies and on the concept of reciprocity is that one has to realize just how technology makes it very difficult for people to talk to each other. People rarely work together on regular, non-technologically interrupted projects. Because of this we have to make the time to create the occasion — be it on the bus, or in the waiting room — to talk to each other not about the weather, but about our "common future."

How do we speak to each other? Here much can be learned from the women's movement — from the way women got together to talk about their status, about the oppression of women historically, politically, and economically. Let us begin with a principled stand and develop a fresh sense of justice. Many of the issues that need to be addressed are best addressed as issues of justice. It can be done by considering simple and everyday things. Look at the size of North American newspapers. Look, for instance, at the *Toronto Star*. Its mere size is a question of justice to me. One needs to ask, "Who has given anyone the right to cut down trees and destroy a habitat for the sake of a double-page advertisement for cars?" These are things that in a caring world cannot be condoned. European papers are small. Europeans still sell cars. There is nothing essential in the magnification of the obvious. Or one can ask, "Who has given the right to publishers to suddenly dish out their newspapers in individual plastic bags that just add to the already unmanageable waste? Who gives the right to owners of large office buildings to keep wasting electricity by leaving the lights on all night in their empty buildings?" These are not questions of economics; they are questions of justice — and we have to address them as such.

You see, if somebody robs a store, it's a crime and the state is all set and ready to nab the criminal. But if somebody steals from the commons and from the future, it's seen as entrepreneurial activity and the state cheers and gives them tax concessions rather than arresting them. We badly need an expanded concept of justice and fairness that takes mortgaging the future into account. Thomas Berger appointed to his inquiry an intervener for the natural environment. But this is a very rare occurrence. The voices of the powerless are not usually heard in technological deliberations, and we have no civic equivalent of the family's practice of "one cuts, the other chooses." If we did, Indian reservations would not be located where they are today.

You may say that the kind of changes required to provide a truly different concept of justice and fairness for decision-making are impossible to achieve. The technological systems — you may say — are so profoundly anchored in our political and social milieu that they cannot be altered so drastically. This I will not accept. There have been profound changes in the past. Slavery was abolished, as was child labour. The status of women has changed quite drastically. All these changes occurred, I suggest, because a point in time came when the general sense of justice and fairness was affronted by, for instance, the owning of people by other people or the exploitation of children, women, or minorities. And to those who may say that slavery was abolished when the institution was no longer necessary in economic terms, that women were liberated when they were required in the workplace, I would say, "Watch it." It seems to me that the sequence of events was likely different. When, on the basis of principled objections, an established social practice has become less and less acceptable, then, and maybe only then, will alternatives be found. From then on tasks that need to be done will be carried out differently and by more acceptable means.

I firmly believe that when we find certain aspects of the real world of technology objectionable we should explore our objections in terms of principle, in terms of justice, fairness, and equality. It may be well to express concerns as questions of principle rather than to try to emphasize merely pragmatic explanations — for instance, that objectionable practices may also be inefficient, inappropriate, or polluting.[10] The emphasis on a pragmatic rationale for choice tends to hide the value judgements involved in particular technological stances.

Today the values of technology have so permeated the public mind that all too frequently what is efficient is seen as the right thing to do. Conversely, when something is perceived to be wrong, it tends to be critiqued in practical terms as being inefficient or counterproductive (a significant term in its own right). The public discourse I am urging here needs to break away from the technological mindset to focus on justice, fairness, and equality in the global sense. Once technological practices are questioned on a principled basis and, if necessary, rejected on that level, new practical ways of doing what needs to be done will evolve.

What I have said may sound like empty words to you, yet it is real and intensely practical. The world of technology is the sum total of what people do. Its redemption can only come from changes in what people, individually and collectively, do or refrain from doing.

Occasions for choice do arise. A decision that I took with respect to my own research may serve as an illustration. Throughout my academic career I have declined to participate in research projects related to atomic energy, which I find an unforgiving and unforgivable technology. My colleagues often suggest that I might still take part in research on nuclear-waste disposal. My response has been a "yes-and-no" one. I have always said that as soon as Canada decides — as Sweden has done — to discontinue the building of nuclear reactors and to phase out existing ones, I would be happy to contribute all I have to addressing problems of nuclear-waste disposal. However, prior to phase-out, effective disposal of nuclear waste is an invitation to produce *more* nuclear waste. I do not wish to have a part in this.

Many of us can make choices and we need to talk to each other about how and why we make these choices. This is an important part of the discourse we need to engage in. This discourse has to be political, and feminist in the sense that the "personal is political." The discourse should be authentic, giving weight and priority to direct experience and reciprocal communication rather than to hearsay or second-hand information. Thus the discourse should seek out those on whom technology impacts.

Attention to the language of the discourse is important. Much clarification can be gained by focusing on language as an expression of values and priorities. Whenever someone talks to you about the benefits and costs of a particular project, don't ask "What benefits?" ask "*Whose* benefits and *whose* costs?" At times it helps to rephrase an observation in line with a perspective from the receiving end of technology. When my colleagues in the field of cold-water engineering speak of "ice-infested waters," I am tempted to think of "rig-infested oceans." Language is a fine barometer of values and priorities. As such it deserves careful attention.

Beyond language, however, the discourse will centre on action — individual as well as collective action. At this point I would like to give two examples of relatively minor changes that could become prototypes for a different interpretation of accountability and reciprocity.

Let me set the scene. We once had a neighbour who was the chairman of the Toronto Transit Commission. Every morning he was conveyed by chauffeur-driven limousine to TTC headquarters, just a few streets north. I felt quite keenly that there was something wrong with this arrangement. Surely those who oversee and guide municipal transportation systems ought to use public transit during their work days. Why not put a clause to that effect in their job description or contract? While the chairman's using public transit would take more time than being chauffeured around, this could become valuable time for learning and reflection.

A job-related constraint, requiring the applicant to use the facilities he or she directs, does not seem to be different in kind from job requirements like speaking the language of the client, agreeing to travel, or not smoking on the job. Requiring those whose work has a major impact on people's lives to experience some of the impact is really not too much to ask. It means that they speak "people" rather than French, Cree, or Spanish.

By the same token, I think that all those who profit from providing food in university cafeterias ought to be compelled by their leases and contracts to have their executives eat that food every day.

The changes in the contract or job specifications advocated in these examples can be accomplished without bringing the country to the brink of bankruptcy or civil war. It would only take the kind of public pressure that comes from political will to use technology — transportation, industrial food preparation, or any other technology — as if people mattered,[11] now and in the future.

To move from the specific to the general: Let's make a checklist to help in the discourse on public decision-making. Should one not ask of any public project or loan whether it: (1) promotes justice; (2) restores reciprocity; (3) confers divisible or indivisible benefits; (4) favours people over machines; (5) whether its strategy maximizes gain or minimizes disaster; (6) whether conservation is favoured over waste; and (7), whether the reversible is favoured over the irreversible? The last item is obviously important. Considering that most projects do not work out as planned, it would be helpful if they proceeded in a way that allowed revision and learning, that is, in small reversible steps.

Making a checklist part of public discourse and expanding the list as needed could be a real help in clarifying and resolving technological and civic issues. However, in the real world of technology there are also situations in which, in fact, one does not know what to do. Henry Regier (former Director of the Institute for Environmental Studies at the University of Toronto) has pointed out that with every development new domains of ignorance are discovered which become evident only as the project proceeds.[12] The emergence of domains of ignorance is basically quite inevitable. Some of the side effects of technical processes could not have been known and are still under study. But the existence of domains of ignorance is itself predictable. This means that it is necessary to proceed with great caution when moving into the unknown and the unknowable. At the same time there have to be resources for adequate research into ways of decreasing the size and depths of these domains. Personally, I think we will soon be faced with a huge domain of ignorance related to the effects of non-ionizing electromagnetic radiation on living organisms. This domain need not have been as big and as deep as it is had there been long-term projects of adequate research paralleling the increasing number of electro-magnetic radiation frequencies.[13]

In addition to the lack of research in the public realm that could diminish the particular domains of ignorance, there is also the lack of recognition of the direct experience of those who work day in and day out in front of video displays or live below high-voltage transmission lines. Their experiences may not be as clear cut and decisive as one might expect to obtain in laboratory research programs, but the initial direct experience of people is an important source of information. To marginalize or discard such direct evidence removes an important source of knowledge from the task of decreasing the domains of ignorance.

But possibly even more important is the implicit attempt to keep people from challenging technology by making their direct experience appear marginal and irrelevant. This is a form of disenfranchisement, and I see disenfranchising people as one of the major obstacles to the formation and implementation of public policies that could safeguard the integrity of people and of nature. This disenfranchising has accelerated since the time of the Industrial Revolution as governments have turned their attention to the blind support of technology and its growth at the expense of other obligations.

The task of redress requires the reintroduction of people into the technological decision-making process. Action is required on a number of levels, including the refusal of consent to our governments to proceed with heroic technology such as nuclear energy, gas and oil mega-projects, or work in space.

In addition, as a country we need to concentrate on the development of *redemptive* technologies, some of which already exist. Others can be developed from different roots. In the first instance, existing technologies need to be reviewed in terms of their scale and the appropriateness of their applications. Initially useful prescriptive technologies are often applied to inappropriate tasks, as when production models and techniques are used in education. Or the scale of a given technology may be the root of its problems, as one finds frequently in agriculture.[14]

Redemptive technologies that arise from the analysis of unacceptable practices of existing technologies should also be accompanied by schemes for assessing appropriateness — like the checklists I just mentioned. New means of technological linkage need to be explored, which would facilitate cooperation without centralization or oppression by scale.

Some redemptive technologies can use existing technical knowledge in a changed structure and for a changed task. For example, I hope that the technical expertise of the Canadian nuclear industry will be redeemed by the industry's providing teams of experts to safety dismantle nuclear reactors around the world when nuclear power becomes globally unacceptable.

In a similar vein, redemptive technologies are needed to prevent pollution. This will mean redesigning industrial processes, reducing waste, and modifying needs and demands. Impressive work of this nature is already in progress. For practical examples see, for instance, Amory Lovins' *Least Cost Energy*.[15]

Another kind of redemptive technology arises from the study of things that *do* work. I have always been amazed that so many resources can be set aside to inquire into things that went wrong. At the same time, few resources are put into the documentation and analysis of processes and institutions that work well — at times in spite of, rather than because of, the system in which they are situated.

Studies of activities and arrangements that work well will be context-specific and carried out largely on the micro level. From such investigations will come knowledge about the factors — usually people with particular gifts — essential for the well-being of the endeavour. New and redemptive technologies can emerge from such studies. When Schumacher spoke about "good work" he argued in support of such technologies.

A third group of redemptive technologies could arise directly from the needs and the experiences of those at the receiving end of the technology. These could address as yet unmet needs — such as personal monitoring of health- and environment-related parameters, easy ways to access relevant information, and low-cost protection of individual privacy against assaults by noise and persuasion. All this would constitute "technology from the bottom up," an approach to new knowledge outlined by Ivan Illich in his essay "Research by people" and by those who speak of "Science and Liberation."[16]

Such bottom-up technologies will incorporate opportunities for reciprocity and include indicators of the onset of problems, related, for instance, to physical, environmental, or institutional health. There could be new and transparent methods for record-keeping and assessment — and much more.

To give just one general example of unmet needs: The field of accountancy and bookkeeping is in urgent need of redemptive technologies. In order to make socially responsible decisions, a community requires three sets of books. One is the customary dollars-and-cents book, but with a clear and discernable column for money saved. The second book relates to people and social impacts. It catalogues the human and community gains and losses as faithfully as the ongoing financial gains and losses documented in the first book. In the third book, environmental accounting is recorded. This is the place to give detailed accounts of the gains and losses in the health and viability of nature, as well as of the built environment.

Decisions on expanding, reducing, or changing particular activities in the real world of technology will require access to and consideration of all three books. Needless to say, adequate technologies of social and environmental accounting do not yet exist; they need to be developed and implemented as part of our search for redemptive technologies.

Finally, the development and use of redemptive technologies ought to be part of the shaping of a new social contract appropriate for the real world of technology, one that overcomes the present disenfranchisement of people. This new social contract needs to be one in which the consent to be governed, regulated, and taxed depends on a demonstrated stewardship for nature and people by those who govern.

I would like to close with an extension of a phrase used by the British peace movement. During their struggle against the placing of Cruise and Pershing missiles on British soil, they urged their fellow citizens to "protest and survive." I want to augment their words and say, "Let us *understand*, and on the basis of our common understanding, *protest*." We must protest until there is change in the structures and practices of the real world of technology, for only then can we hope to survive as a global community.

If such basic changes cannot be accomplished, the house that technology built will be nothing more than an unlivable techno-dump.

VII

As I revisit, in 1999, the house that technology built, I find it difficult to single out those features that have changed most dramatically during the past decade. Much of what I observed then — the dominance of prescriptive over holistic technologies, the growing culture of compliance and its civic impacts — remains prominent. Yet in terms of "the way things are done" the new electronic technologies left their deepest impacts on society's perceptions of time and space and on the way we, as citizens, relate to each other. This is why I have chosen to focus in these new chapters on issues of communication, of perceptions of time and space, and on collaboration.

Let me begin my reflections on communication technologies with a simple scheme of definitions, which I hope will make the subsequent discussions of the impact of modern technologies more transparent.

"Communications," as I will use the term, is essentially the transmittal of a message from a sender to a receiver: The basic process is direct and unmediated, normally one-to-one and of the "I love you" or "you still owe me ten bucks" type that can be illustrated schematically by s —> r. One sender can, of course, send a verbal message to a large number of receivers, as in the case of a sermon, a lecture, or a proclamation: s —> R.

A message can also be sent directly by a large number of senders to one receiver, such as when the crowd before Pontius Pilate shouted "Barabbas," or when demonstrators chant in unison "No, no, we won't go." The case is then S —> r.

As these examples illustrate, the communication mode is always direct, immediate, and simultaneous. Everyone involved is present in the same place throughout the process of communicating.

The first extension from the s —> r mode comes with the introduction of a messenger. As the sender entrusts the message to an intermediary, the range of communication can be extended, both in time and in space. Though the physical presence of the messenger still links sender and receiver, the integrity of the message now demands assurance of the truthfulness of the messenger, or communication turns into gossip: s —> m —> r.

The profound impact of writing, as a technology, lies in the fact that writing allows the physical separation of the message from the messenger or sender. When Hammurabi set down the details of Babylon's laws or

Moses proclaimed the ten commandments, the "writ" became the message that would survive the writer/sender and could be brought to distant places.

Beyond reflecting on the impact of writing on the preservation of culture, custom, and thought, a related facet of this impact may deserve mention here: the possible relationship of writing to orthodoxy and fundamentalism. This is how I perceive the connection:

Let us distinguish for a moment those cultures that rely on writing for the transmission of their moral and religious values, such as "the people of the book" — adherents of Judaism, Christianity, or Islam — from those cultures who entrusted their most important social, religious, and cosmological knowledge to elders and priests. Oral traditions do not codify transmission and interpretation of laws and values in the way that textually based societies do. Indeed, strict adherence to the letter of the law is only possible when the letter or the writ exists and carries with it the belief in its authority almost regardless of context. And herein might lie, I think, a major root of fundamentalism and some traditions have warned against overreliance on "the letter": in 1656, Quaker Elders concluded an epistle with the words ". . . that these things may be fulfilled in the Spirit, not from the letter, for the letter killeth, but the Spirit giveth life."[1]

Seen as a new technology, a new way of doing things, writing extends the range of communication both in time and in physical space. This is why we know so much more of, and are influenced so much more by, past civilizations that put their thoughts and accounts in writing than by those who did not. Not that historical accounts necessarily tell the truth or even convey the message as intended by the sender, but wherever they exist, ancient texts appear incredibly authoritative, often solely by virtue of their antiquity.

However, since the very beginning of writing there have been attempts to assure the authenticity of a given communication. Seals or signatures, detailed identifications of the writer or the source of the material — all these add-ons have attempted to substantiate the integrity of the message. Writing by hand, it seems, extended the range of receivers more than the range of senders, but it was printing and the related increase in literacy that truly increased the circle of both senders and receivers. In terms of receivers this led, for instance, to a much greater diffusion of technical and scientific knowledge.[2] The new presses could print pamphlets and broadsheets for senders who expressed views and comments intended to change the social order. Christopher Hill's *The World Turned Upside Down* chronicles the social movements in sixteenth- and seventeenth-century England that could not have come into being without printing technology.[3]

As the variety of messages increased due to the increasing spectrum of senders and receivers, so grew the need to further authenticate, verify, and classify the messages. This task began to require the specific identification of the sender and/or writer as well as an assurance of the veracity of the message's content. In terms of communication, the channel of transmission through which a sender's message reached the receiver — I will call it the conduit —

became a significant factor in the authentication process. The imprimatur of the publisher, the reputation and standing of editors and compilers became, to their contemporaries, symbols of the acceptability and of the political or moral thrust of a text. Thus both content and intent of the message would be assessed from the publisher's reputation; the attributes of the conduit came to signal the character of the message. This trend encouraged senders and distributors to identify their motives and social positions. (Sixteenth- and seventeenth-century British Quakers, for instance, called themselves "Publishers of the Truth.")

Through books, tracts, pamphlets, and newspapers, communication activities increasingly became part of public life. While these communications can be looked upon as coming from more or less identifiable senders, it is no longer possible to clearly identify their receivers or even to assume that the complete message has been received. Indeed, many such communications have to be regarded as messages looking for receivers. Communication-related social inventions, such as imprimatur, the characterization of authors, background references, quotations, and footnotes, should be interpreted as aids that assist the receiver in assessing the messages offered. As well, the growth of libraries and archives made comparison and cross-reference possible and necessary. All these subchannels of communication are part and parcel of the pre-electronic communication endeavour, brought about by the growth and diffusion of writing and printing technologies. With such aids, receivers and users of communications in print could learn to "read between the lines."

The use of electrical and electronic technologies significantly altered the nature of communications. As discussed earlier, in chapter two, these changes have all occurred within the past 150 years, although the major systemic changes are even younger. All of them began slowly. The use of the telegraph, as one of the early electrical communication devices, resulted in greater speed for the transmission of messages and allowed greater physical distances to be bridged in acceptable times, but senders and receivers remained clearly identifiable. In terms of the basic communication scheme, the original telegraph was mainly a modification of the conduit. Initially largely in the hands of the powerful, the mode of telegraphic message transmission, with its awkward mechanics of sending and receiving, made this conduit best suited for the sending of directives or commands.

The telephone brought genuine change to the existing communication processes. Through the use of the phone, sender and receiver can be in direct contact without being physically in each other's presence. While sender and receiver are separated in space, their communication is direct and unmediated by a messenger. Their conversation can be as private as to permit a sender to say "I love you" or "you still owe me ten bucks" without anyone else hearing it. The technology eliminates the need for sender and receiver to be in the same place, but, by requiring their simultaneous presence on the line, the telephone restores the possibility of directly transmitting an authentic message. The influence of the conduit on the content is exercised mainly through the cost of phoning.

I have spoken earlier about the reciprocity that true communication requires, of the back and forth of speaking, listening, and then supplying a response informed by what has been heard.[4] Although the phone eliminates from the dialogue the body language that face-to-face exchange includes, the direct telephone conversation can still contain genuine reciprocity. Nevertheless, one must not forget that the phone can also facilitate misrepresentation and fraudulent schemes.

Once the scientific knowledge of the electrical and electronic properties of matter began to be translated into devices and components for communication — i.e. the sending and receiving of messages — the impact of electronic technologies became quite comparable to those of writing.

Electronic technologies have wrought a number of changes that are important for us to note here. One is the increased speed and range of the transmission of signals. Then there is the ability of electronic recording techniques to separate sound from its source and — as part of the same technology — to recombine and create sound, whose origin can be no longer identified by the listener.[5]

The ability to transmit images — by means of digitizing them — and to recombine the transmitted bits permits us to send messages containing text, sound, and images to those who are technically equipped to receive them. Finally, and maybe most importantly, the new electronic technologies — into which I count computers and their networks — allow the collection, storage, redirection, and retrieval of vast amounts of information. And here I quite deliberately use the term "information" for the first time. In many ways the spread of electronic transmission of bits of information marks the end of the simple notion of communication as the transfer of a message from a sender to a receiver. The notion of information, defined as "knowledge communicated concerning some particular subject, or event"[6] places the emphasis on content and format, rather than intent.

There are a number of ways in which the new real world of information technology is different from the traditional world of communications, sketched above. The differences are much more profound than the extension of the existing systems, such as e-mail and fax supplanting the old-fashioned postal service for the transmittal of messages. I have touched on some of these aspects in the earlier lectures, particularly in chapter two. Let me therefore move directly to computer systems and to the Internet, the major symbol and playing field of the new information technologies.

The Internet,[7] the omnipresent decentralized global system of interconnected computer networks, was originally designed to fulfill the military's need for the instant and resilient exchange of data. The decentralized design, which allows access from all sites on the Net to all sites of the Net, lies at the core of its impact and has played a central part in the revolutionary changes that computer technologies have brought. It is essential to remember that computers not only compute, manipulate numbers, and store data for fast retrieval, but they also transform information — words, sounds, and images — into bits that can be transmitted in digitized form and be received and recombined into words, sounds, or images at the point of request. Combining the potential of digitization with the properties of computer networks is in many ways like inventing writing again, though now in very new dimensions — and I use dimensions in the broadest possible sense of the world.

In my analogy, computers know a new alphabet called bits. When you and I write down a verbal message, we use phonetics to put into words what we hear; our transcript can be kept, sent, or read by those who know our language. Electronic devices can digitize and render into bits the information before them. A string of bits can be stored, transmitted, and reconstructed just as a verbal message can be, through writing and reading.

In a historical setting a scribe would create words from sound, put them in a code of language on clay tablets or papyrus, and send them off. Another scribe, familiar with the code, reads the text to the designated receiver. I see computers as today's scribes who, by means of digitizing (their form of writing, or code), make bits from my writing and e-mail them to my friends' e-mail boxes in the computers, which present my letter to them within seconds of my computer having sent it. The new scribes know how to read and write in "BIT" and have learned how to translate text, sound, or images into BIT and BIT back into images, sound, or text. BIT then becomes a language just like Hebrew or Arabic, not just a technical term. Computers writing in BIT bring a new dimension to communications that I consider equivalent to that of writing. I am aware that regarding BIT as a language raises a multitude of scholarly questions. Because language is part of our being human, the designation "language" carries profound cognitive and perceptual implications. I am not unaware of these, but I still would like to use the term "language," or "idiom," in the more colloquial sense, because it expresses so well the link between BIT and communication.

The pathways for transmitting messages between computers are mostly outside the human landscape; computer messages move in cyberspace, that nonhuman realm or domain into which our electronic communications have been transposed. Thus cyberspace has become the virtual or nonhuman and unreal environment for the transfer, storage, and retrieval of messages in BIT. I have never liked the term cyberspace because it neither describes a space nor does its current use reflect the concepts of control and systems-design implied in the term cybernetics, after which the term cyberspace was patterned.[8] I use the designation cyberspace reluctantly, and only because I need a name for the realm that is unreachable by human feet and yet is of ever increasing import to human life.

Subsequent chapters will examine some of the interfaces between cyberspace and the sphere of human life. These interfaces demonstrate the discontinuities that the new technologies, the new ways of doing things, have brought to the world's human, social, and political activities. For the balance of this chapter, though, I want to explore some implications of writing in BIT for the enterprise of communication in the more narrow sense of the term.

If one accepts the image of computers as the new scribes writing in BIT, there are then analogies between the problems of communicating over computer networks and those encountered during the earlier phases of written and printed communications. There are again the questions of authentication and verification of messages, the identification of sender, receiver, or transmitter, the detection of the origin and intent of the information transmitted, not to mention issues of privacy or ownership of content. The professional practices that evolved as part of writing and printing technologies — from imprimatur and editorship to the "reading between the lines" — are being revisited today and sometimes reinvented. But this is not a simple undertaking, because we are not dealing here merely with recasting an old task — that of sending and receiving messages — into a new technological setting. We have to deal with different and quite new social relationships that now superimpose existing ones. As well, cyberspace has become not only the new channel of transmission, but also a new realm of storage, assembly, and distribution of information.

I got into real trouble once, when I suggested that the Internet could be looked at as one giant dump: people and organizations dump information in bits and pieces; they also retrieve whatever is of use and interest to them. What is found by the scavengers depends on where they dig, what was dumped, and what is considered useful or relevant enough to be retrieved. There is no pattern in the digging or reassembly, no one path through the dump, no compulsory reference to the scource of the bounty. And since the Internet contains information rather than stuff, the same treasures, or junk, can be retrieved again and again. My colleagues felt that my dump image did not do justice to the great equalizing value of the open structure of the Net, nor to the nonhierarchial nature of its access and the significance of the vast information resources available to everyone on the Net. They may well be right. What attracted me to the dump image was that it conveyed the lack of any link between those who dump and those who retrieve, as well as the absence of a means of verification or attribution of prior ownership.

My interest is mainly to understand the changes that all the new technologies have brought and to comprehend their impacts. The changes that the new information technologies have brought influence the individual mind, but they also structure our social and political institutions. While the changes appear abrupt and even catastrophic,[9] they must be seen as continuations of preceding techno-logical developments, which have tried to change the constraints that time and space put on human pursuits. More on this in the next chapter.

VIII

In the previous chapter, I tried to systematize my under-standing of "communication" and indicate the impact of the new technologies — from writing to printing and computers — on the practice of sending and receiving messages. I began to illustrate the effects of the changing practices on the content and scope as well as on the pathways of such communications. I will continue this exploration here.

The use of transparent language and the need to choose appropriate words poses a difficult task for me. How do we speak about the new? As human beings, we communicate with each other by referring to common experiences: we try to compare and contrast. The new has to be expressed in a comparison with the old as "more so," "less so," "like," or "unlike." Still, there may be more to the new than just being novel and so one looks for analogies, for metaphors. At times, inappropriate metaphors are coined intentionally so as to mislead (see chapter four); but unsuitable metaphors may also arise from a lack of appreciation of the relevant attributes of the new. The term "information highway" may come to mind if one thinks of regulation and infrastructure, but the human reality of our transposed communication activities is probably more pointedly expressed in one of my favourite cartoons. In it a receptionist is sitting in a library. Behind her are rows of readers, each in front of a computer screen. She is answering a patron's inquiry with "Sir, this is a library. If you want a book, go to a bookstore." As the artist succinctly questions the traditional assumption that libraries are full of books, we are stimulated to be mindful of the many ways in which new technologies change established patterns, not all of them easily expressed in words.

Many technological innovations have been introduced in order to change the boundaries of human and social activities with respect to time and space. In many ways, time and space are the two sides of the coin of human existence. Whatever changes one side will affect the other, and often the terms "faster" or "further" are quite inadequate to describe their impacts. This is especially true in any discussion of the new computer-based network technologies, and I would like to stress some more telling features of time and temporality before examining further the network issues of cyberspace.

But while I now need to look more closely into the scholarship of human existence in space and time, I want to emphasize my reason for doing so: I would like to understand better the human, social, and ecological impact of technology; i.e., the way things are done. The deeper questions of time, space, and the meaning of meaning I must leave to philosophers and theologians. There is no contribution that I can make to their disciplines. I just worry about what happens to people, to their communities, their culture, their land, and their future under the weight of the new practices. I want to understand, as well as I can, this new real world of technology, because that is where all of us live. And this is why I pick at the available wisdom, looking for some thoughts that might illuminate the problems that have arisen and that will have to be coped with.

Nature provided human societies with their earliest markers of time. These markers range from the positions of stars, the sun and moon, and the rhythm of the seasons to the migration patterns of animals, and the birth, growth, and decay of all living things. From these experiences comes the understanding of patterns and cycles and the place of human societies within them. But there is a clear distinction to be made between measured time and experienced time.

Time is at the centre of people's personal and collective sense of identity, which in turn is based on a shared history, on a common knowledge of the sequence of relevant past events. The sacred books of many civilizations recount such events in terms of their sequence: ". . . in the beginning there was . . ." In the same vein, individual lives and their stories have a beginning and an end. Time is always with us, we speak of giving time, making time, wasting, stealing, and even killing time. Women speak of "that time of the month." There is due time for a child to be born. In all this, time is *real* and it patterns human existence as it structures our collective and personal memory. It is well to remember that Immanuel Kant saw time and space not as external media within which people move, but as ordering devices of the human mind.

Throughout history, human beings have rattled the cages of space and time that confined their natural existence. Human inventions have perfected the reckoning of elapsed time, be it by means of the controlled flow of water, by reading the shadow's position on a sun dial, or later mechanical devices, and are a testimony to the bonding and bondage of society and time. Lamps pushed back the night, domestication and crossbreeding of plants extended the growing seasons. To evade the inevitable consequences of nature's seasonal patterns, nations fought for ice-free sea ports and built empires "upon which the sun never sets." But even when new technologies, from telegraph to railways, bridged great distances in short times, the newly created international time zones acknowledged the reality of nature's predominance over time. "The Clock Carol" a song by Donald Swann, written in 1965, resonates this reality in its opening line: "When the bells chime noon in London, New York begins its day, 'good morning' in Toronto means 'good night' in Mandalay."[1]

One has to remember also that the very preoccupation of technology with outwitting the constraints of time and space is, in and of itself, evidence of the profound grip of temporality on society. It is the facet of time as sequence and pattern that, in terms of the new technologies, requires our attention most. The concept of *synchronicity* and its opposite, *asynchronicity*, are central to this aspect of temporality and technology.

Synchronicity will be used here in the meaning that Carl G. Jung[2] and Lewis Mumford[3] gave it. Mumford stressed how the use of clocks as the instruments for marking time into commonly acknowledged segments changed the structure of communities as well as that of individual work and living. The bell's call to work or prayer keeps a community "in sync," often imposing more and more detailed patterns of dominance on individuals and groups.

Jung, on the other hand, drew attention to the importance of common patterns and acausal coincidences in establishing a sense of meaning and connectedness within and between individuals. Thus, while synchronicity evokes the presence of sequences and patterns, fixed intervals or periodicities, coordination and synchronization, asynchronicity indicates the decoupling of activities from their functional time or space patterns.

I have already discussed the history of the prescriptive repatterning of work and social relations through technological manipulations of time in factory, school, or prison (see chapter 3). Most of these developments can be interpreted as powerful new patterns that structured people's lives by imposing new time and space configurations.

However, the current widespread use of computer networks and related technologies has led to something different: the prevalence of asynchronicity, indicated by the loosening, if not the abandonment, of previously compulsory time and space patterns.

This is a most significant change. No longer is one pattern superseded by another pattern; the change now appears as a move from an existing pattern to no discernable structure. I consider the evolving destructuring by asynchronicity as an extremely important, if not *the* crucial facet of the new electronic technologies.

The role of asynchronicity in unravelling social and political patterns without apparent replacement with other patterns cannot be overestimated. Let me give you just a few examples of asynchronicity and its social and human consequences. In terms of communication, take voicemail, as we have come to call it, even though the "voice" is produced by a device, and the "mail" may never be delivered. The ping-pong pattern of verbal communication is no longer tied to space or time. You can send your ping to someone before going to lunch; she may pick it up when and where she can and, at some point, send back her pong — and there goes the joy of intimate contact, and with it the heart of what verbal communication was thought to be: an exchange of messages between people in the present tense. The point here is not so much the interposing of devices into a process of communication, but the changing of a synchronous process into an asynchronous one.

Many people have experienced the asynchronous forms of labour and have felt their consequences; the impact often includes the lack of work-related solidarity and self-identification that can have profound social implications. And then there are the visits at all hours to the libraries of the Internet, the electronic articles and books that do not have to be returned. There are the new haves and have-nots, now defined in terms of their ownership of equipment, their access to and knowledge of the new codes that allow asynchronic practices.

What does this all mean to us as humans, as social and political beings, evolving within the patterns of nature and culture? It's not that I believe everything asynchronous is "bad" and that everything synchronous is "good." Not at all. Women in particular have often treasured the opportunity to work asynchronously — getting a bit of writing done when the kids are asleep, sneaking in a slice of private life into their tightly structured existences. But I see a real difference between *supplementing* a rigidly patterned structure with asynchronous activities and *substituting* synchronous functions by asynchronous schemes. I will elaborate on these distinctions later because what troubles me is not as much the nature of asynchronous processes but their increasing prevalance, if not dominance.

But let me now turn to a book entitled *City of Bits*,[4] published in 1995, that exquisitely illustrates this dominance. The author, William Mitchell, juxtaposes the old realities of time and space with the new bit-world of asynchronicity. Mitchell takes an architectural approach to living and working asynchronically. In the City of Bits the e-mail address becomes one's domicile, the Internet the always open library, the chat-group the extended family of choice, the World Wide Web the "information flea market," as Mitchell calls it. Work and play intertwine to the point of nonseparation, and social justice is defined in terms of access to cyberspace for business, pleasure, and politics. A brilliantly thoughtful and frightening book, it is informed by the research and outlook of the MIT Media Labs and gives the reader a scary sense of inevitability as to the coming of a global "bitsphere." For Mitchell the issue is not whether the world will turn into a global bitsphere, but how the design of the bitsphere is evolving.

Mitchell's vision is as breathtaking for what it depicts as for what it does not. The inhabitants of the City of Bits are still real live human beings, yet nature, of which humans are but a small part, appears to have no autonomous space in the bitsphere. There are no seasonal rhythms, no presence of the land nor the ebb and flow of individual lives, even though these are the synchronous patterns that have shaped culture and community throughout the time and, through their patterns, have provided a source of meaning to people for many generations.

To substitute, rather than supplement, these synchronous patterns with the asynchronous practices of the bitsphere is not a trivial matter to contemplate. The biosphere, the living earth, people within their communities, their culture, and their history are all still with us. The built environments of the past, the cities and towns, the roads and waterways were all constructed sequentially, one item after another. The sense of history and identity, present in every civilization, is rooted in a common knowledge of past events and their time sequences. History, after all, is as someone once said, "just one damned thing after another." Human beings perceive life in its physical, social, and political dimensions as having evolved in steps and stages, not as being assembled and constructed from distinct pieces according to a specific design or blueprint. Sequence and consequence are intimately connected in the human mind; can one let go of sequence and maintain the notion of consequence, let alone accountability?

Before discussing the borders between the bitsphere and the human sphere and the struggles for dominance taking place between the synchronous and the asynchronous approaches, I need to bring two concepts, two different realities, into our discussion. We need to remember the difference between a *mechanism* and an *organism*. This is not a new distinction, but an extremely important one in the context of the real world of technology. Brian Goodwin's formulation, stated recently in the context of his discussion of evolutionary biology, has been of great help to me. Going back to Kant, Goodwin writes:

> Kant described a mechanism as a functional unity, in which the parts exist for one another in the performance of a particular function. The clock was the paradigmatic machine in his time. Pre-existing parts, designed to play specific roles in the clock, are assembled together into a functional unity whose dynamic action serves to keep track of the passage of time.
>
> An organism, on the other hand, is a functional *and* structural unity in which the parts exist for *and by means of* one another in the expression of a particular nature.

This means that the parts of an organism — leaves, roots, flowers, limbs, eyes, heart, brain — are not made independently and then assembled, as in a machine, but arise as a result of interactions within the developing organism.[5]

In light of these definitions, it is tempting to see Mitchell's bitsphere as the grand and ultimate mechanism receiving and delivering messages between senders and receivers in an asynchronous manner through cyberspace. The bitsphere seems to hover over the multitude of diverse organisms — physical or social — and appears to devour them one by one. It is a seductive image but, I think, not necessarily an appropriate one. Rather than "being there" in the sense of being a mechanism in the broader sense of the term, the structures and functions in the bitsphere are assembled for a specific task or transaction only and may or may not exist beyond the life of the transaction. For an analogy of the operation of the bitsphere, you may want to think of music on a page; it does not "sound" until played. Although I know musicians who can "hear" a score when looking at it, generally music has to be played or sung to be heard by others. The bitsphere, like a musical score, has to be accessed to come to life. And it is the differential access to the bitsphere and its potential that pulls so strongly on the world's social and political fabric.

As I see it, the bitsphere is not a supermechanism but rather a nonorganic environment designed to be accessed and utilized asynchronically. This means that connected computers can transmit messages not only at any time and from or to any place, but that cyberspace, as a conduit, does not impose a pattern or sequence on the transmittal or content itself, beyond the possibility of asynchronic use.

Having centred my attention in this chapter on the changing relationships to time that are permeating the world of the new technologies, I now need to look at the other side of the coin of human existence: the changing relationships of people and physical space.

IX

In chapter eight I introduced the *City of Bits*, William Mitchell's discussion of the architecture of living and working in a world structured by the asynchronicity of the new electronic environments, environments he called the bitsphere. One of my comments had been that human beings, their communities and histories, as well as the biosphere and its complexity do still exist. Fascination with new technologies can change the focus of our perception of what is actual and real. Sometimes I have to remind myself that after all one does not bump into a website when walking the dog, though one may meet the neighbours and talk about the potholes in the street.

In this chapter I would like to illuminate the struggles at the interface between the human sphere and the bitsphere. To help us, I have cooked up an image that allows me to illustrate the structural and historical dimensions of the problems at hand.[1]

Imagine the whole world as a round cake; its wedge-shaped slices are states or countries. As residents of a slice, we are closer to adjacent slices than to more distant ones. Within the slice we can picture social mobility as a vertical structuring, rearrangements of place between, say, the crumb of the bottom and the icing on the top, with the raisins in between. Community, then, is locality, as is its representation. Democracy has local roots, its first practice is local: the member of Parliament from Kicking Horse Path represents a small slice within the larger slice called Canada.

Throughout history, language, law, and custom have been identified vertically in terms of locale; locale has been slice dependent. The very notion of foreign languages or of a lingua franca is an acknowledgement of the local nature of language, culture, and community. Yet slices were rarely completely isolated; exchanges across the cuts — whether state borders or boundaries of language or ethnicity — have always existed, more likely, of course, between adjacent slices than between distant ones. Throughout history people have travelled across large distances to foreign lands. They have returned home with new impressions, knowledge, and goods. One can consider this traffic of individuals, goods, and ideas as the horizontal slicing of the global cake.

For a long time the extent of such horizontal movements was mitigated by geography and was regulated by local, or vertical, laws. Instruments such as passports, tariff and border surveillance, protected vertical activities from the encroachment of horizontal endeavour. Many technological innovations, intended to push against the constraints of time and space, have made it easier to achieve horizontal movements: from navigational instruments that improved sea travel to railways, cars, and airplanes, from telegraph and radio to telephone, fax, and e-mail — all have contributed to a vast increase in the cross-border and cross-country exchanges. Add to this the fact that the modern production technologies, with their prescriptive fragmentation, lend themselves very well to global subcontracting and asynchronistic reassembly and you get a feeling for the globalization blueprint.

We all know specific realities of the increasing dominance of horizontal activities: goods and parts are manufactured overseas for sale here; services are offered by voices identified by first names only from call centres located who knows where; our newcasts contain stock market news from around the world as frequently as they bring local traffic reports. The instant updates on the routes money travels around the world appear as important to Canadian citizens as hearing about a car accident that may delay their trip home from work.

This state of things has not arrived overnight. The growing importance of horizontal transactions demanded facilitation and regulation, beginning with customs unions, relaxing of visa requirements, and the easing up of travel and currency regulations. But the quantum jump came again through the electronic technologies. In terms of the cake model, within the vertical slices, some parts benefitted much more than others from these developments, often, though not always, because of their place within the slice.

One has to realize that the laws intended to regulate the effects of the horizontal activities on the vertical patterns, have to be made by countries, i.e. by the very states whose capacity to regulate and structure the vertical patterns would be impeded by the creation of autonomous horizontal patterns. Such processes usually result in a divestment of powers of the vertical in favour of the horizontal. As, for instance, when nation states surrender their regulatory authority in certain areas to transnational corporations, or when international trade agreements take precedence over national law. The ruling apparatus of a country often splits into a horizontal fraction, interested in easing the horizontal activities and therefore attempting to decrease the cohesion and power of the local, or vertical, component, concerned with the conditions within the slice. In Canada, the social and political consequences of, for instance, the North American Free Trade Agreement (NAFTA) or the Multilateral Agreement on Investment (MAI) illustrate the tug of power between the horizontal and the vertical forces.

The changes in the Canadian tax system provide a further illustration of the effect of the horizontal and the vertical forces on the governance of our country. In 1955, 43 percent of the income tax collected by the governments of Canada came from corporations, the rest from personal income taxes. In 1995 the corporations' share was 11 percent.[2] From the point of view of the government, the rationale for such differential treatment is that, while the citizens are stuck in the slice, corporations can slither along the horizontal cuts in search of a suitable global tax haven. From the point of view of the citizen, the issue is one of representation, of democracy: they ask, who do the elected members that make the laws and regulations or give consent to their implementation actually represent? Their vertical slice, or the forceful horizontal cut through it?

Another facet of globalization that the cake model illustrates well is the current predominance of investment capitalism over manufacturing or production capitalism.[3] In chapter one I introduced the concept of prescriptive technologies. These practices began during the Industrial Revolution in mid-eighteenth–century Europe. Using machines and a new division of labour, the new technologies greatly increased the production of material goods. Within Western Europe — and later North America — they restructured work and gave new opportunities for the concentration of power and capital. In the then emerging technological order, the relationship between government and business changed and, as I pointed out using the distribution of electricity as an example, the state began to establish the infrastructures that enabled the subsequent growth of commercial production, while at the same time encouraging consumption and materials acquisition. This general trend has continued, but its scope and range has become enlarged and augmented, basically by increasingly wide-ranging financial activities. What has happened can be seen again as the interplay between new technologies and the consequences of the applications of the old ones.

The mass production of consumer goods, even in highly automated facilities and low-wage and -overhead locations, did not yield the boundless harvest of profit that some had expected. In part this is due to the fact that those who most need the mass produced goods — and this includes food, medicine, and clothing — do not have the means to purchase the very items they often make.[4] At the same time, those who have disposable incomes have become more discriminating, as well as somewhat satiated. All in all, consumerism isn't what it used to be.

On the other hand, the bitsphere, with its new computer technologies of command and control, greatly facilitates fast and asynchronous monetary transactions. This in and of itself has led to a very substantial increase in global financial trading and profit making. The sphere of speculation and investment has grown with the growth of the bitsphere. While the production of goods has its end-roots somewhere in a locality, somewhere in a vertical slice, money and investment funds are no longer anchored. Note, for instance, that the trading floor of the Toronto Stock Exchange no longer exists as a physical space, but only in the bitsphere. Yet the transactions at the Exchange are real. They bring gains or losses to real people, offer hope to some, despair to others, largely by virtue of opaque horizontal transactions globally conducted in BIT.

The new dominance of investment capitalism over production capitalism can be seen well in the changed relationship between business and the nation state: when the acquisition of raw materials and the sale of material goods was the dominant form of profit making, the state — as I indicated before — took an active role in providing a political climate conducive to commercial endeavours, by employing instruments of regulation, and public policy making. Thus, citizens were regarded mainly as consumers and were encouraged to act as such. The shift of profit making from direct production to investment has not changed the active support of commerce by the state, but it has drastically changed its support strategies. These changes have led to a remarkable shift in the relationship between the state and its citizens, a shift not often commented upon.

Given the global economy, it became evident to business in the industrialized parts of the world that major profits could be made if the public sector were opened up for private investment. Historically, a country's not-for-profit sector, from roads and parks to schools, hospitals, and jails, is firmly set into its respective vertical slice. Many public institutions are structured and regulated with local requirements and shared values in mind, not the least of them the democratic conviction that the needs of some — such as children, the elderly, or the infirm — are not a source of profit for others. Yet the recent privatization of many public sector functions and the deregulation of their operations have meant that the government has opened the vertical, community-rooted tasks — traditionally entrusted to the oversight of the state — to the horizontal forces of global investment. Thus the state has delivered the most dependent of its citizens, as well as resources held in trust, as new investment opportunities to the global market.

The question here is not whether individuals are better served by the private or public sphere; the issue, I would argue, is one of governance and responsibility. The destruction of the public sector and the loss of the coherence of the vertical through the pull of the horizontal has profound effects on community and on the bonds between people. Changes in this area should not be initiated without thorough public discussion of the broader issues at stake.

Not everything that is technologically possible, such as bitsphere-driven investment capitalism, is either desirable or necessary for the well-being of a country. I regret that public and local discussions on the impact of the new technologies on life in Canada have had so little impact on the governance of our country. I fear that our slice of cake is crumbling without our consent.

So much for the cake model. I hope it has served to illustrate that the real word of technology, within which we try to live and work in peace, with a modicum of justice and equality, is in the grip of strong and competing forces. The cake model can be linked readily to some of our earlier discussions. Asynchronicity is an essential feature of working along the horizontal cuts, while synchronicity and shared patterns provide much of the cohesion of the vertical. Clearly, each and every one of us is affected by both synchronistic and asynchronistic practices.

In the final chapter I will examine some facets of living and working in the increasingly asynchronistic environment that constitutes the borderland between the bitsphere and the biosphere.

X

In chapter seven I examined the domain of communication, from ancient to modern, and introduced computers as the new scribes that could translate language, sound, and image into the new idiom of BIT; I tried to present the impact of digitization and computer networking in terms of analogies or correspondences to the invention of writing. In chapters eight and nine I focused on the dimensions of time and space, respectively, and reflected on how the new technologies, the new ways of doing things, have altered the time-space context within which we all live and work.

In this last chapter I would like to put some of these thoughts together and look at the interfaces between the bitsphere and the biosphere, the realm in which most of today's world exists. Here the term biosphere will be used in its broadest sense to include not only all living creatures and their biological support systems, but also the physical and mental artifacts attesting to their presence on earth (you may want to think of a nest of spheres, embedded in each other). The bitsphere is seen as the sphere of storage, display, and transmittal of information or data in BIT.

Individually and collectively we live simultaneously in the bitsphere and the biosphere, sometimes more profoundly in one, sometimes more intensely in the other. But this is the reality of our time, whether we like it or not, and no one can make one sphere or the other go away. All one can do is to try and understand, as well as possible, the current and potential dynamics of these interacting spheres of influences and try to monitor and mitigate their impacts, remembering Kant's insight that time and space are ordering devices of the human mind.

I will focus on three interrelated facets of our lives — education, work, and governance — to illustrate the impacts, but many others could be used. The resulting insights, I hope, will be applicable to other situations because of the common dynamics of pushing and pulling between the bitsphere and the biosphere.

As I compile some characteristics of the biosphere and the bitsphere so as to compare and contrast their attributes, I do so not to create an either/or picture, but to illustrate the changing milieu that the *simultaneous* presence of conflicting and often incompatible forces generates. Within the biosphere, human beings have attempted to codify and transmit their understanding of the world around them by ordering their experiences into general schemes and structures.[1] Myths, religion, and science have endeavoured to transmit knowledge and experience so ordered as to convey sequence and consequence as ordering principles. Learning to recognize such ordering principles has been traditionally part of growing up in a given society. Ordering schemes help us to evaluate and interpret new knowledge and experience. Ordering schemes, in turn, are revised and augmented by new insights. The very concept of order, sequence, and correlation is drawn from the observation of nature.

It is salutary to remember nature's inescapable sequences, to not forget that the seed has to be in the soil before growth can begin, that the trees do not grow into the sky, the day has no more than twenty-four hours, and that the sun rises and sets regardless of political events. Those concerned with the advancement of technology, with finding new and more encompassing ways of doing things, often underemphasize the nonnegotiable embeddedness of human society in nature. No one, individually or collectively, can separate from the biosphere. You just can't opt out of nature.

In addition to being embedded in nature, people and their social and political groupings are nested in their particular cultures, as formed by the interplay of actions, thoughts, and values of past generations. While the character of social and political institutions has changed through time, social structures *per se* have persisted and provided ordering principles to define culture and practice.[2] Thus individuals carry the imprints of social structures just as they carry nature's imprints.

One of the most striking attributes of the bitsphere, on the other hand, is the absence of structure. As far as I can understand it, the bitsphere is designed to have no structure that would correspond to the coherent sequence/consequence patterns of past conduits of messages. Conceptually, one can get around this apparent absence of predictive structures of the bitsphere by evoking higher levels of complexity[3] or utilizing the insights of chaos theory.[4] However, for my purpose of understanding the social and political impacts of the new technologies, it seems more helpful to stress the anti-structuring aspects of the bitsphere, i.e. the asynchronistic and inherently fragmented practices. While the biosphere, existing in real time, encompasses past, present, and future, the bitsphere — a product of human minds — exhibits no tense or temporality, and no roots in physical space. As an environment for transactions, the bitsphere uses the stored or imported data solely for transmittal on demand and does so most efficiently.

Where does this leave us, as we consider education, work, and governance at the interface of bitsphere and biosphere? Earlier, I contrasted a growth model of education, rooted in an organic context, with a production model, based on mechanistic considerations. In the latter, schooling is regarded as a production process that can be analyzed and evaluated in terms of input, output, efficiency, and cost effectiveness. During the past decade, this model has dominated educational politics, and educational technologies were developed and introduced accordingly. The hope that the new communications technologies, particularly film and radio, television, and computers would broaden and deepen the scope of education, just as writing and printing had done in the past, has been largely realized.

Unfortunately, the new technologies have entered the realm of education largely because they were regarded as production improvements, promising better products and faster or bigger production runs, and not because they were deemed to offer enrichment to the soil. Thus it is not surprising that the electronic classroom raises the same types of problems and exhibits the same social and political difficulties that one encounters in the realm of work or governance in the real world of the new technologies.

As I see it, these common problems and difficulties fall basically into three streams: the first occurs because of the displacement of people by devices,[5] an extension of the Industrial Revolution's old dream of the workerless factory to the school without teachers, or the government without officials.

The second stream, arising as a consequence of the first, contains the problems caused by the vast underestimation of the contributions to a task made by those working or learning together,[6] as well as a lack of appreciation of the fact that knowledge is cumulative.

The third stream contains the problems created by the increase of asynchronistic modes of doing things and the resulting social time-space dislocations.[7]

The daily realities, of course, encompass problems from all three streams, each being influenced by the presence of the others.

Let me stay with the area of education for a moment and look at some of the dynamics of teaching and learning. Whenever a group of people is learning something together, two separate facets of the process should be distinguished: the explicit learning of, say, how to multiply and divide or to conjugate French verbs, and the implicit learning, the social teaching, for which the activity of learning together provides the setting. It is here that students acquire social understanding and coping skills, ranging from listening, tolerance, and cooperation to patience, trust, or anger management. In a traditional setting, most implicit learning occurred "by the way" as groups worked together. The achievement of implicit learning is usually taken for granted once the explicit task has been accomplished. This is no longer a valid assumption. When external devices are used to diminish the need for the drill of explicit learning, the occasion for implicit learning may also diminish.

Yet the implicitly learned social skills and insights may be much needed, even when the explicit skills can be obtained externally via appropriate devices such as spell checks, calculators, or computers. Without an adequate understanding of the social processes of teaching and learning and a careful attention to their well-being, the whole enterprise of education can be at risk.[8]

I have time for just one illustration, one image of what the substitution of a device for an activity can do. When I first came to Canada as a post-doctoral fellow, I was surprised by the severity of the injuries that my colleagues incurred while downhill skiing. It took me a while to understand the situation. Although I had skied before coming to Canada, I was not familiar with ski lifts. It had been my experience that once one managed to get up a hillside, one had acquired enough skill to get down reasonably safely. The ski lift removes the "by the way" opportunity to learn how to climb, fall, and get up again, as well as the ongoing reality check on fitness and resources. If such knowledge is not present before using the lift, there is real danger to life and limb. Once understood, it is relatively simple to gain the experience needed to use the ski lift safely.

When transferring this analogy to the classroom, it is not that simple to spot the lost "by the way" opportunities for social learning and then to create replacements for them. The production model of education makes it difficult to even acknowledge the existence of implicit learning, let alone compensate for its loss. After all, the new technologies were intended to bring efficiency and cost savings to education as a production activity. Thus, many of the teachers who could help to reshape the new pedagogical settings so as to preserve and encourage implicit learning become surplus. Neither their presence nor their experience appears any longer to be a necessary part of the process of education.

As considerations of efficiency and cost-cutting shift the balance of synchronous and asynchronous classroom activities, the balance of explicit and implicit learning is changing. While the pool of information available to the students may increase, the pool of available understanding may not. This has considerable consequences for social cohesion and peace and deserves careful attention.

The situation in the classroom at the interface between the biosphere and the bitsphere is but one facet of the situation in the workplace within the same realm. In fact, often even the designation of workplace is no longer appropriate. Not only do new technologies, new ways of doing things, eliminate specific tasks and workplaces — telephone operators and their switchboards, statisticians and their files — but the remaining work is frequently done asynchronously in terms of both time and space.

But how and where, we ask again, is discernment, trust, and collaboration learned, experience and caution passed on, when people no longer work, build, create, and learn together or share sequence and consequence in the course of a common task?

The prescriptive technologies that I described at the beginning of this book displaced many holistic technologies and led to the fragmentation of work as well as its increasing specialization. The automation of production that David Noble and others have documented[9] prepared the way for the globalization of manufacturing that today so displaces labour and dislocates work in terms of time and space. The existence of the bitsphere enables an unprecedented acceleration of the deconstruction of traditional work patterns. When people no longer work together in the same place — the shop floor, the typing pool, the warehouse or the factory — opportunities for social interactions, for social learning and community building disappear, just as the implicit learning opportunities in the classroom can vanish when the cohesion of learning in a group is eclipsed by the device-assisted, individually-paced acquisition of knowledge. But where, if not in school and workplace, is society built and changed?

The impact of the space-time dislocations of work on individuals and their surroundings has attracted some serious attention recently.[10] But what about the substance of work itself? Work provides livelihood as well as meaning, identity, and purpose and, as we live at the interface of biosphere and bitsphere, all these components are under the impact of the new technologies.

I think Fritz Schumacher beautifully articulates the place of work in our lives when he writes that

> . . . we may derive the three purposes of human work as follows:
>
> First, to provide necessary and useful goods and services.
>
> Second, to enable every one of us to use and thereby perfect our gifts like good stewards.
>
> Third, to do so in service to, and in cooperation with, others, so as to liberate ourselves from our inborn egocentricity.[11]

"This threefold function," Schumacher continues, "makes work so central to human life that it is truly impossible to conceive of life at the human level without work. 'Without work, all life goes rotten,' said Albert Camus, 'but when work is soulless, life stifles and dies.'"

Work, in Schumacher's perspective, is in and of the biosphere. His notion of working is a far cry from the activities of the agile, flexible, and disposable labour force that the globalized economy favours.[12] You will say, surely, that work is more than gainful employment; we work in our homes, our gardens, we make music together, or serve on committees. Human fulfillment may come from work that is unrelated to income-producing jobs. Yet all of us need nourishment for body and soul and the interface between the bitsphere and the biosphere is a risky place for both body and soul.

In a technologically structured world, individuals are rarely able to assure their physical and emotional survival solely by their own wits. What might increase their chances for a full and healthy life are the social, political, and economic structures of their country. The viability of such structures, characterized in the previous chapter as belonging to the vertical slice in the cake model, depends to a large extent on the viability of the country as a whole and its responsiveness to change. What, then, of the role of the state in this turbulent border realm of biosphere and bitsphere?

In chapter nine I used the cake model to illustrate the tensions on governance that the pull of the horizontal forces has asserted. The inability, but mostly the unwillingness, of the state to intervene in order to limit the social and human impacts of the new technologies on their own citizens has been documented thoroughly. This "retreat from governance" as H. T. Wilson calls it,[13] is relatively new. It had long been assumed that being a citizen, belonging to a nation or a community, conferred a measure of practical and emotional security. But at the interface of biosphere and bitsphere, the reality of togetherness and belonging becomes eroded by the asynchronous activities in virtual time and space. As the nation state bows to the forces of local and global commerce, vital social and human structures become deeply eroded.

Today in Canada, the practice of democratic governance is in grave question and the advancement of social justice and equality appears stalled in a labyrinth of random transactions.[14] This does not have to be so. The interface of the biosphere and the bitsphere not only poses problems and precipitates crises but it offers new opportunities to advance the common good. It will take the collective thought, moral clarity, and strong political will of many people to move towards this goal rather than away from it.

CODA

What, then, can we do, trapped as we are in this historical situation, that appears to me so unstable and threatening?

Much of what I said at the end of my 1989 Massey Lecturers (chapter six) remains valid. Justice and peace, reciprocity and community matter as much as (or more than) ever and so does the language of our discourse about the real world of technology. Over and above this, we need to bring into this discourse the contradictory attributes of the bitsphere and the biosphere, so as to understand more clearly the impact of the changing balance of synchronistic and asynchronistic processes. For it is here that the new practices of the past ten years have revealed most acutely new danger to individuals and communities but also new possibilities of human and social progress.

Again, I plead for thorough communication, clarity, and

collective action, and I rejoice in the many new activities in this area.

Here the bitsphere offers unprecedented opportunities for exchanges of information and coordinated interventions. Such initiatives permit the voicing of opposition where no parliamentary opposition is possible. There have been remarkable results of international human rights interventions, responses to environmental emergencies, and in breaking official information monopolies on projected international agreements, such as the MAI.[1] However, at present, the successes are situational, not yet systemic.

One precondition for pressing for systemic changes is an understanding of the ongoing dynamics of technology and power. This is why I always stress the need for clarity and understanding of the realities of the world of technology however complex they may appear. We can help each other to see things that are commonly not placed in the political foreground: For instance, over the unending din of economic rhetoric, we need to speak of what happens to people.[2] What happens to people is not a mere footnote to an economic report, but should be the central focus for action of governments and communities.

There has been a phenomenal lack of clarity and concern on the part of governments of Canada regarding the fate of people and communities caught between the bitsphere and the biosphere. An example of this lack is the absence of any serious public discussion on social mitigation, such as a basic income policy[3] or the taxation of electronic transactions — the bit tax or Tobin tax.[4] If and when such discussions occur, great respect for the structures of society and for the biosphere is essential. The "by the way" contributions of our nested environments to social peace and understanding are probably much greater than presently acknowledged. Often one can appreciate the importance of such contributions only when they are at risk or lost.

For a long time the technologically induced changes of the dominant time and space practices have been exceptions within the ongoing traditional patterns. The balance of life was still anchored in the structured rhythms of the social realm and existed consciously within the biosphere. The increasing inroads of the bitsphere and its asynchronous practices requires us, I think, to be much clearer and much more articulate about the essential contributions of synchronicity to our well-being; we must press for their retention. We may not know precisely how much asynchronicity a person can tolerate before being socially and morally dislocated. We may not know all about the factors that stabilize or destabilize communities, although we know a fair amount: poverty and lack of meaningful work destroys, as does lack of respect, friendship, and reciprocal commitments. As I see it, asynchronous practices can both help and destroy meaning and connectedness. The crucial difference is whether the asynchronous processes supplement the synchronous practices or are a substitute for them.

A Toronto civic organization, Citizens for Local Democracy (C4LD for short) will illustrate my point. Born in the effort to oppose the regional amalgamation of the five local municipalities, it became a lively focus for political action and discernment. Its strength has been built with a remarkable mix of communications: information is disseminated simultaneously by e-mail and print, discussed on the Net, and in regular meetings; there is a website and a phone tree. The combination has built a community of common purpose where there was none before.[5] Only time can give us an assessment of the advancement of common goals that this and other similarly structured democratic efforts could attain.

In the next few years, much will have to be learned and shared, particularly with respect to the nature and actions of governance. During the past decade, the house that technology built has become larger and more unified, though, I have argued, neither more livable nor more beautiful. In 1990 the Royal Society of Canada published a book called *Planet Under Stress: The Challenge of Global Change*[6] to illuminate the impact of human activities on the global ecosystem. In these new chapters of *The Real World of Technology* I have tried to show how deeply integrated people are into the biosphere and I could have well called the new edition *Communities Under Stress: Is Being Human Threatened by Global Change?* The impact of human activities on all peoples as well as on nature needs to be central in any forthcoming political discourse. Societies are organisms that evolve through and with their members; they are not mechanisms to be assembled and disassembled at will. Yet the potential of doing just that exists to date on a global scale because of the characteristics of asynchronistic technologies. Thus the struggle to understand and steer the interaction between the bitsphere and the biosphere is the struggle for community in the broadest ecological context.

This is a collective endeavour that no group or conglomerate can do on its own. Most of our social and political institutions are both reluctant and ill-equipped to advance such tasks. Yet if sane and healthy communities are to grow and prevail, much more weight has to be placed on maintaining the non-negotiable ties of all people to the biosphere.

NOTES

Chapter I

1. C. B. Macpherson, *The Real World of Democracy* (Toronto: Canadian Broadcasting Corporation, 1965); C. B. Macpherson, *Democratic Theory: Essays in Retrieval* (Oxford: Clarendon Press, 1973).

2. For a general survey of the field, see Paul Durbin, ed., *A Guide to the Culture of Science, Technology, and Medicine* (New York: The Free Press, 1980), particularly Part I, chapter 2; Part II, chapter 5; and Part III, chapter 7. For special aspects, see P. L. Bereano, ed., *Technology as a Social and Political Phenomenon* (New York: John Wiley, 1976); Stephen Hill, *The Tragedy of Technology* (London: Pluto Press, 1988); Donald MacKenzie and Judy Wajcman, eds., *The Social Shaping of Technology: How the Refrigerator Got its Hum* (Milton Keynes: Open University Press, 1985); Joan Rothschild, ed., *Machina Ex*

Dea: Feminist Perspectives on Technology (Toronto: Pergamon Press, 1983).

3. For definitions of technology, see Carl Mitcham, "Philosophy of technology," in Paul Durbin, ed., *A Guide to the Culture of Science, Technology, and Medicine*; George Grant, "Knowing and making," in *Royal Society of Canada Proceedings and Transactions*, 4th series, 12:59-67, 1974.

4. The quoted definition is from Kenneth E. Boulding, "Technology and the changing social order," in David Popenoe, ed., *The Urban-Industrial Frontier* (New Brunswick, NJ: Rutgers University Press, 1969). See also Kenneth E. Boulding, *The Image* (Ann Arbor: University of Michigan Press, 1956); Kenneth E. Boulding and L. Senesh, eds., *The Optimum Utilization of Knowledge* (Boulder, CO: Westview Press, 1983); R. P. Beilock, ed., *Beasts, Ballades and Bouldingisms* (New Brunswick, NJ: Transaction Books, 1980); Kenneth E. Boulding, *Three Faces of Power* (London: Sage Publications, 1989). For technology as practice, see also Max Weber, *The Theory of Social and Economic Organization* (Oxford: Oxford University Press, 1947).

5. David F. Noble, "Present tense technology" in *Democracy*, Spring/Summer/Fall, 1983. Reprinted as a monograph, *Surviving Automation Madness* (San Pedro, CA: Singlejacks Books, 1985).

6. Ursula M. Franklin, "The beginning of metallurgy in China, a comparative approach," in G. Kuwayama, ed., *The Great Bronze Age of China: A Symposium* (Seattle: University of Washington Press, 1983).

7. Vita Sackville-West, "The Land," from *Collected Poems* (London: The Hogarth Press, 1934).

8. M. J. Herskovits, *Economic Anthropology* (New York: Alfred

A. Knopf, 1952).

9. For a general discussion on the division of labour and the factory system, see Christopher Hill, *Reformation to Industrial Revolution*, volume 2 of *The Pelican Economic History of Britain* (New York: Penguin, 1967); Eric Hobsbawm, *Industry and Empire: From 1750 to the Present Day*, volume 3 of *The Pelican, Economic History of Britain* (Baltimore: Penguin, 1977); Andre Gorz, ed., *The Division of Labour* (Hassocks: Harvester, 1978).

10. D. P. S. Peacock, *Pottery in the Roman World* (London: Longman, 1982). For Roman technology in general, see K. D. White, *Greek and Roman Technology* (Ithaca, NY: Cornell University Press, 1984).

11. Ursula M. Franklin, "On bronze and other metals in early China," in D. N. Kneightley, ed., *The Origins of Chinese Civilization* (Berkeley: University of California Press, 1983). See also Franklin, "The beginning of metallurgy in China," and P. Meyers and L. Holmes, "Technical studies of ancient Chinese bronzes," in G. Kuwayama, ed., *The Great Chinese Bronze Age of China*.

12. Ursula M. Franklin, J. Berthrong, and A. Chan, "Metallurgy, cosmology and knowledge: The Chinese experience," *Journal of Chinese Philosophy*, 12:4, 1985.

13. Langdon Winner, *The Whale and the Reactor: A Search for Limits in an Age of High Technology* (Chicago: University of Chicago Press, 1986).

14. The notion of externalities and the related concept of total costing is discussed extensively, particularly in relation to technological assessment. For a general perspective, see *Canada as a Conserver Society* (Ottawa: Science Council of Canada, Report #27, 1977).

15. Maxine Berg, *The Machinery Question and the Making of Political Economy, 1815-1848* (Cambridge: Cambridge University Press, 1980).

16. Ursula M. Franklin, "Where are the machine demographers?," *Science Forum*, 9:3, 1976.

17. R. P. Beilock, ed., *Beasts, Ballades and Bouldingisms* (New Brunswick, NJ: Transaction Books, 1980).

Chapter II

1. The word "vernacular" is not used here exactly in the sense that it is woven into the work of Ivan Illich; however, I chose to use this term in order to emphasize the resonance with Illich's thought, for instance in his essay "Vernacular values" in *Shadow Work* (Boston: M. Boyars, 1981).

2. For background information on feminism, see, for instance, Jessie Bernard, *The Female World* (New York: Free Press, 1981); Marilyn French, *Beyond Power* (New York: Summit Books, 1985).

3. W. H. Vanderburg, *The Growth of Minds and Cultures* (Toronto: University of Toronto Press, 1985) and "The John Abrams Memorial Lectures," in *Man-Environment Systems* (special issue on science, culture, and technology) 16:2/3, 1986.

4. M. L. Benston, "Feminism and the critique of scientific method," in *Feminism in Canada*, A. Miles and G. Finn, eds. (Montreal: Black Rose Books, 1982); H. Rose and S. Rose, eds., *Ideology of/in Natural Sciences* (Cambridge, MA: Schenkman, 1980).

5. B. J. Bledstein, *The Culture of Professionalism* (New York: Norton, 1976); Barbara Ehrenreich and Dierdre English, *For Her Own Good: 150 Years of the Experts Advice to Women*

(London: Pluto Press, 1979); Carolyn Marvin, *When Old Technologies Were New* (Oxford: Oxford University Press, 1988). W. Armytage, *The Rise of the Technocrats: A Social History* (London: Routledge and Kegan Paul, 1965); David Collingridge, *The Social Control Of Technology* (New York: St. Martin's Press, 1980); M. J. Mulkay, *Science and the Sociology of Knowledge* (London: G. Allen & Unwin, 1979); J. R. Ravetz, *Scientific Knowledge and Its Social Problems* (Clarendon: Oxford University Press, 1971).

6. Cheris Kramarae, ed., *Technology and Women's Voices* (New York: Routledge and Kegan Paul, 1988); Heather Menzies, *Fast Forward and Out of Control* (Toronto: Macmillan of Canada, 1989); Ruth Schwartz Cowan, "The Consumption function," in *The Social Construction of Technological Systems*, W. E. Bijker, T. P. Hughes and T. J. Pinch, eds., (Cambridge, MA: MIT Press, 1987); Donna Smyth, "The Citizen scientists — What she did not learn in school," in *Canadian Women's Studies*, 5:4, 1984.

7. Jerry Mander, *Four Arguments for the Elimination of Television* (New York: Morrow, 1978); Paul Goodman, *New Reformation: Notes of a Neolithic Conservative*, (New York: Random House, 1970); Stewart Brand, *The Media Labs: Inventing the Future at MIT* (New York: Viking, 1987); Elisabeth Noelle-Neumann, *The Spiral of Silence: Public Opinion, Our Social Skin* (Chicago: University of Chicago Press, 1984); Todd Gitlin, ed., *Watching Television* (New York: Pantheon Books, 1986); Mark Crispin Miller, *Boxed In: The Culture of TV* (Evanston, IL: Northwestern University Press, 1988); Eric Pooley, "Grins, gore and videotape," in *New York* magazine, pp. 75-83, 9 October 1989. For a self-defence textbook prepared under the aegis of the Association for Media Literacy, see Barry Duncan, *Mass Media*

and Popular Culture (Toronto: Harcourt Brace, 1988).

8. World Commission on Environment and Development, *Our Common Future* (Oxford: Oxford University Press, 1987).

Chapter III

1. Heather Menzies, *Fast Forward and Out of Control* (Toronto: Macmillan of Canada, 1989); Donald MacKenzie and Judy Wajcman, eds., *The Social Shaping of Technology: How the Refrigerator Got its Hum* (Milton Keynes: Open University Press, 1985).

2. *The Science and Praxis of Complexity*; contributions to the symposium held at Montpellier, France, 9-11 May 1984 (Tokyo: United Nations University, 1985).

3. Jacques Ellul, *The Technological System*, translated by J. Neugroschel (New York: Continuum, 1980). See also: W. H. Vanderburg, *Perspectives on Our Age: Jacques Ellul Speaks on His Life and Work*, translated by J. Neugroschel (Toronto: CBC Enterprises, 1981).

4. Within the extensive literature on systems, these two books show the classic foundations and the conceptual developments: L. von Bertalanffy, *General System Theory* (New York: George Braziller, 1968); Anatol Rapoport, *General System Theory* (Tunbridge Wells: Abacus Press, 1986).

5. Pam McAllister, ed., *Reweaving the Web of Life: Feminism and Nonviolence* (Philadelphia: New Society Publishers, 1982); Vandana Shiva, ed., *Staying Alive* (London: Zed Books, 1988).

6. Michel Foucault, *Discipline and Punish*, translated by A. Sheridan (New York: Vintage Books, 1979).

7. C. Babbage, *On the Economy of Machinery and Manufactures* (London: C. Knight, 1832).

8. David F. Noble, "Present tense technology," in *Democracy*, Spring/Summer/Fall, 1983. Reprinted as a monograph *Surviving Automation Madness*.

9. William Petty, "Of the growth of the city of London," in *The Economic Writings of Sir William Petty*, vol. 2, 1682, C. H. Hull, ed., (Cambridge: The University Press, 1899).

10. Maxine Berg, *The Machinery Question and the Making of Political Economy*, 1815-1848 (Cambridge: Cambridge University Press, 1980).

11. Robert Owen, *A View of Society and Report to the County of Lanark*, V. A. G. Gattrell, ed., (London, 1962); Robert Owen, *The Life of Robert Owen, Written by Himself* (New York: A. M. Kelley Publishers, 1967); Sidney Pollard and J. Salt, eds., *Robert Owen, Prophet of the Poor* (London: Macmillan, 1971).

12. Lewis Mumford, *The Condition of Man* (New York: Harcourt, Brace and Company, 1944).

13. David S. Landes, *The Unbound Prometheus: Technological Changes and Industrial Development in Western Europe from 1750 to the Present* (Cambridge: Cambridge University Press, 1969); Paul McKay, *Electric Empire: The Inside Story of Ontario Hydro* (Toronto: Between the Lines Press, 1983); Siegfried Giedion, *Mechanization Takes Command* (New York: Norton, 1969).

14. Thomas P. Hughes, *Networks of Power: Electrification in Western Society*, 1880-1930 (Baltimore: Johns Hopkins University Press, 1983).

15. This decision constituted an effective gate against other suppliers of electricity (see McKay above) and off-shore suppliers of appliances and machine tools. It should be noted that for the last twenty years new residences in Ontario have been wired for 220-volt dual phase power to accommodate

appliances such as clothes dryers. However, this does not mean that any European gadget can be used in a Canadian home without modification. These points of clarification arose from correspondence with Mr. D. J. C. Phillipson.

16. Robert A. Caro, *The Power Broker: Robert Moses and the Fall of New York* (New York: Alfred A. Knopf, 1974).

Chapter IV

1. E. F. Schumacher, *Small Is Beautiful; Economics as if People Mattered* (New York: Harper & Row, 1973); George McRobie, *Small Is Possible* (London: Jonathan Cape, 1981); Hazel Henderson, *Creating Alternative Futures* (New York: Berkeley Publishing Group, 1978).

2. Here I am referring to an event that happened in the spring of 1985. The Minister of Energy, Mines and Resources informed me in a letter dated 15 April 1985 that I had been appointed to a two-year term on the five-member Atomic Energy Control Board. However, within days the Prime Minister's Office announced that my appointment had not occurred and that the minister's letter was sent due to an administrative error. For an account, see *Hansard*, vol. 128, nos. 90, 91, 92, 1985.

3. From the large body of writing on international peace, I have chosen a few examples that focus on the fundamental and structural aspects of the issues: Kathleen Lonsdale, *Removing the Causes of War: The Swarthmore Lecture, 1953* (London: Allen and Unwin, 1953); American Friends Service Committee, *In Place of War; An Inquiry into Nonviolent National Defense* (New York: Grossman Publishers, 1967); Suzanne Gowan, *Moving Toward a New Society* (Philadelphia: Society Press,

1976); *Nuclear Peace*, a CBC *Ideas* program by Ursula Franklin, Jan Fedorowicz, David Cayley, and Max Allen (CBC Enterprises broadcast transcript, Toronto, 1982); Johan Galtung, *There are Alternatives!: Four Roads to Peace and Security* (Nottingham: Spokesman, 1984); Seymour Melman, *The Demilitarized Society* (Montreal: Harvest House, 1988).

4. Anatol Rapoport, "The technological imperative," in *Man-Environment Systems* 16:2/3, 1986.

5. For historical roots, see Mulford Q. Sibley, *The Quiet Battle* (New York: Anchor Press/Doubleday, 1963). For the current Canadian situation, see *Conscience Canada Newsletter*, Victoria, B.C.

6. Berit Ås, "On female culture," in *Acta Sociologia* (Oslo), vol. 2-4:142-161, 1975.

7. Thomas Berger, *Northern Frontiers, Northern Homeland*, report of the Mackenzie Valley Pipeline Inquiry, vols. 1-2 (Ottawa: Supply and Services Canada, 1977).

8. *Canada as a Conserver Society* (Ottawa: Science Council of Canada, Report #27, 1977).

9. Carolyn Merchant, *The Death of Nature* (San Francisco: Harper & Row, 1980); Barry Commoner, *The Closing Circle* (New York: Alfred A. Knopf, 1971).

10. Nigel Calder, ed., *The World in 1984*, vols. 1-2 (Harmondsworth: Penguin Books, 1964).

Chapter V

1. See ch. 2, ref. 1.

2. Herbert Marcuse, *One Dimensional Man* (Boston: Beacon Press, 1964).

3. Humphrey Jennings, *Pandaemonium: The Coming of the Machine*

as Seen by Contemporary Observers, 1660-1886 (London: A. Deutsch, 1985).

4. Marshall McLuhan, *The Mechanical Bride* (New York: Vanguard Press, 1951).

5. Dennis Gabor, *Innovations: Scientific, Technological, and Social* (Oxford: Oxford University Press, 1970); Tracy Kidder, *The Soul of a New Machine* (London: Allen Lane, 1982).

6. See ch. 2, ref. 6.

7. Cheris Kramarae, "Lessons from the history of the sewing machine," in *Technology and Women's Voices*, Cheris Kramarae, ed. (London: Routledge and Kegan Paul, 1987).

8. Joan Rothschild, ed., *Machina Ex Dea: Feminist Perspectives on Technology* (Toronto: Pergamon Press, 1983); Sandra Harding, *The Science Question in Feminism* (Ithaca, NY: Cornell University Press, 1986); Evelyn Fox Keller, *Reflections on Gender and Science* (New Haven, CT: Yale University Press, 1985); Cynthia Cockburn, *Machinery of Dominance* (London: Pluto Press, 1985).

9. Karin D. Knorr-Cetina and Michael Mulkin, eds., *Science Observed: Perspectives in the Social Study of Science* (London: Sage Books, 1983); Sally Hacker, *Doing It the Hard Way: Essays on Gender and Technology*, Dorothy Smith and Susan M. Turner, eds. (Boston: Unwin Hyman, 1990); C. DeBresson, M. L. Benston, and J. Vorst, eds., *Work and New Technologies: Other Perspectives* (Toronto: Between the Lines Press, 1987); Sally Hacker, *Pleasure, Power and Technology: Some Tales of Gender, Engineering, and the Cooperative Workplace* (Boston: Unwin Hyman, 1989).

10. Ursula M. Franklin, "Will women change technology or will technology change women?," in *Knowledge Reconsidered:*

A Feminist Overview, selected papers from the 1984 annual conference (Ottawa: Canadian Research Institute for the Advancement of Women, 1984).

11. Friedrich Klemm, *A History of Western Technology* (New York: Scribners, 1959); Ivy Pinchbeck, *Women Workers and the Industrial Revolution, 1750-1850* (London: Routledge and Sons, 1930; reprinted by A. M. Kelley, New York, 1969).

12. See ch. 2, ref. 5.

13. Norbert Wiener, *The Human Use of Human Beings*, 2nd edition (New York: Anchor Press/Doubleday, 1954).

14. Elaine Bernard, "Science, technology and progress: Lessons from the history of the typewriter," *Canadian Women's Studies*, 5:4, 1984.

15. Helen Potrebenko, *Life, Love and Unions* (Vancouver: Lazara Publishers, 1987).

Chapter VI

1. John Murra, *The Economic Organization of the Inka State* (Greenwich, CT: JAI Press, 1980). For more information on the context and meaning of this setting, see, for instance, Sally Falk Moore, *Power and Property in Inca Peru* (New York: Columbia University Press, 1958).

2. Ilya Prigogine and Isabelle Stengers, *Order Out of Chaos: Man's New Dialogue with Nature* (Boulder, CO: New Science Library, 1984).

3. C. S. Holling, "Perceiving and managing the complexity of ecological systems," in *The Science and Praxis of Complexity*; contributions to the symposium held at Montpellier, France, 9-11 May 1984 (Tokyo: United Nations University, 1985).

4. Marcus G. Raskin and Herbert J. Bernstein, *New Ways of*

Knowing: The Sciences, Society, and Reconstructive Knowledge (Totawa, NJ: Rowman & Littlefield, 1987); Sandra Harding and Merril Hintikka, *Discovering Reality: Feminist Perspectives on Epistemology, Metaphysics, Methodology, and Philosophy of Science* (Boston: D. Reidel, 1983).

5. See, for instance, Etienne Balazs, *Chinese Civilization and Bureaucracy*, translated by H. M. Wright (New Haven, CT: Yale University Press, 1964); Alberto Guerreiro Ramos, *The Science of Organizations: A Reconceptualization of the Wealth of Nations* (Toronto: University of Toronto Press, 1981). The scope of these lectures does not allow time for reflections on Max Weber's thoughts on rationalization, though they are at the root of many later deliberations; for an introduction, see, for instance, Hans Gerth and C. Wright Mills, eds., *From Max Weber: Essays in Sociology* (Oxford: Oxford University Press, 1946).

6. See ch. 4, ref. 7.

7. See ch. 4, ref. 8.

8. E. F. Schumacher, "Technology for a democratic society," included in George McRobie, *Small Is Possible* (London: Jonathan Cape, 1981).

9. David Dickson, *Alternative Technologies and the Politics of Technical Change* (London: Fontana/Collins, 1974); Karl Hess, *Community Technology* (New York: Harper & Row, 1979); E. F. Schumacher, *Good Work* (New York: Harper & Row, 1979); Arnold Pacey, *The Culture of Technology* (Oxford: Blackwell, 1983).

10. For discourse along these lines, see, for instance, Gregory Baum and Duncan Cameron, *Ethics and Economics* (Toronto: Lorimer, 1984); Daniel Drache and Duncan Cameron, *The Other Macdonald Report* (Toronto: Lorimer, 1985).

11. Ursula M. Franklin, "New approaches to understanding technology," in *Proceedings of the International Seminar on Technology, Innovation and Social Change* (Ottawa: Carlton University Press, 1984; reprinted in *Man-Environment Systems* 16:2/3, 1986); W. H. Vanderburg, "Political imagination in a technological age" in *Democratic Theory and Technological Society*, R. B. Day, Ronald Beiner, and Joseph Masciulli, eds. (Armonk, NY: M. E. Sharpe, 1988).

12. Henry A. Regier, "Will we ever get ahead of the problems?," in *Aquatic Toxicology and Water Quality Management*, J. A. Nriagu, ed. (New York: John Wiley, 1989).

13. See, for instance, Paul Brodeur, *Currents of Death* (New York: Simon and Schuster, 1989).

14. Brewster Kneen, *From Hand to Mouth: Understanding the Food System* (Toronto: NC Press, 1989).

15. Amory Lovins et al., *Least Cost Energy: Solving the CO_2 Problem*, 2nd ed. (Snowmass, CO: Rocky Mountain Institute, 1989).

16. Ivan Illich, "Research by people," in *Shadow Work*; R. Arditti, P. Brennan, and S. Caviak, eds., *Science and Liberation* (Montreal: Black Rose Books, 1980).

Chapter VII

1. "Christian faith and practice in the experience of the Religious Society of Friends," London Yearly Meeting of the Religious Society of Frends (Quakers) 1960.

2. Stillman Drake, *Mechanics in Sixteenth-Century Italy* (Madison: U of Wisconsin P, 1969).

3. Christopher Hill, *The World Turned Upside Down* (London: Penguin, 1975).

4. See chapter 2, p. 48.

5. For a more detailed discussion of this aspect see Ursula M. Franklin, "Silence and the notion of the commons," in *Procedings of the Conference in Acoustic Ecology* (Banff 1993). A short version of the lecture was published in *Musicworks 59 (The Journal of Sound Exploration)* (Summer 1994).

6. *Oxford English Dictionary*, 2nd ed. (Oxford: Claredon Press, 1989).

7. For the origin of the Internet, see, for instance, Jeffrey A. Hart et al., "The Building of the Internet," *Telecommunications Policy* (November 1992), 666–89.

8. "Cyberspace" began as a literary term in William Gibson's *Neuromancer* (New York: Avon Books, 1984). The concept of cybernetics originated with Wiener and Rosenbleuth in 1947. See Norbert Wiener, *Cybernetic* (Cambridge, MA: MIT Press, 1969).

9. William Greider, *One World, Ready or Not* (New York: Simon & Schuster, 1997). Michel Chossoudrovsky, *The Globalisation of Poverty* (Atlantic Highlands, NJ: Zen Books, 1997).

Chapter VIII

1. Donald Swann, *Sing round the Year* (London: The Bodley Head Ltd., 1965) 68.

2. Carl G. Jung, *Memories, Dreams, Reflection* (New York: Random House, 1961), 221, 388. See also Jung, "Synchronicity, an acausal connecting principle," in *Carl Jung: Collected Works*, Vol. 8 (London: Routledge and Kegan Paul, 1954). See also Victor Mansfield, *Synchronicity, Science and Soul-Making* (Chicago: Open Court Publishing Co., 1995).

3. Lewis Mumford, *Technics and Human Development* (New York: Harcourt Brace Jovanovich, 1966), 286.

4. William J. Mitchell, *City of Bits: Space, Time and the Infobahn* (Cambridge MA: MIT Press, 1995).

5. Brian Goodwin, *How the Leopard Changed His Spots: The Evolution of Complexity* (New York, London: Simon & Schuster, 1994). Here again I have to admit that I cannot do justice to the depth and richness of this area of inquiry. Many thinkers have pointed out that the notions of mechanical and organic models, in spite of their utility, leave out essential human and social dimensions that encompass both the personal and the spiritual. These aspects of life are profoundly affected by changing technologies, but neither the scope of these lectures nor my own scholarship allows me to address them adequately. I can only hope that others will do so.

Chapter IX

1. Ursula M. Franklin, "Beyond the hype, thinking about the information highway," *Leadership in Health Services*, (Ottawa: CHA, July/August 1996). Franklin, *Every Tool Shapes the Task, Communities and the Information Highway* (Vancouver: Lazara Press, 1996).

2. Walter Stewart, *Dismantling the State* (Toronto: Stoddart Publishing, 1998), 294-5.

3. See chapter 7, note 9 and John Dillon, *Turning the Tide: Confronting the Money Traders* (Ottawa: CCPA, 1997).

4. Consult, for instance, the monthly magazine *New Internationalist* for reports on world poverty and inequality, and the reports of the relevant agencies of the United Nations.

Chapter X

1. Kenneth E. Boulding, *The Image* (Ann Arbor: U of Michigan, 1956). With respect to science, see also David Knight, *Ordering the World: A History of Classifying Man* (London: Burnett Books, 1981).

2. Bruce G. Trigger, *Sociocultural Evolution* (Oxford: Blackwell Publishers, 1998).

3. *The Science and Practice of Complexity* (Tokyo, Japan: The United Nations University, 1985).

4. For an introduction to chaos thoery see, for instance, Ian Stewart, *Does God Play Dice?* (London: Penguin, 1990).

5. David F. Noble, *Progress without People* (Toronto: Between the Lines, 1995). Jeremy Rifkin, *The End of Work, the Decline of the Global Laborforce and the Dawn of the Post-Market Era* (New York: G. P. Putnam, 1995).

6. For an appreciation of the contributions of workers to the workplace technologies, see for instance DeBresson *Understanding Technological Change* (Montreal: Black Rose Books, 1987); Karen Messing, *One-eyed Science, Occupational Health and Women Workers* (Philadelphia, PA: Temple UP, 1998); as well as the work of David F. Noble.

7. Heather Menzies, *Whose Brave New World?* (Toronto: Between the Lines, 1996). Richard Sennett, *The Corrosion of Character* (New York: W. W. Norton & Co., 1998).

8. Ursula M. Franklin, "Personally happy and publicly useful" in *Our Schools / Our Selves* 9.4 (October 1998).

9. David F. Noble, *Forces of Production: A Social History of Industrial Automation* (New York: Knopf, 1984). Rifkin, *Timewars* (New York: H. Holt, 1987). Rifkin, *The End of Work*.

10. See for instance Armine Yalnizyan, T. Ran Ide, and Arthur

Cordell *Shifting Time: Social Policy and the Future of Work* (Toronto: Between the Lines, 1994).

11. Ernst Friedrich Schumacher, *Good Work* (New York: Harper & Row, 1979).

12. Heather Menzies, *Whose Brave New World?* (Toronto: Between the Lines, 1996). Jamie Swift, *Wheel of Fortune: Work and Life in the Age of Falling Expectatons* (Toronto: Between the Lines, 1995).

13. Linda McQuaig, *The Cult of Impotence: Selling the Myth of Powerlessness in the Global Economy* (Toronto: Viking, 1998); Stewart, *Dismantling the State*; and an early and important warning regarding Canada was given by H. T. Wilson, *Retreat from Governance* (Hull, QC: Voyageur Publishing, 1989).

14. For details see note 13 (above), and, for instance, Tony Clarke, *Silent Coup* (Ottawa: Canadian Centre for Policy Alternatives and Lorimer, 1998). For more global and philsophical perspectives see Maria Mies, and Vandana Shiva, *Ecofeminism* (London and New Jersey: Zen Books, 1993).

Coda

1. Gregory Albo and Chris Roberts, "The MAI and the world economy" in *Dismantling Democracy: The Multilateral Agreement on Investment (MAI) and It's Impact* (Ottawa: Canadian Centre for Policy Alternatives and Lorimer, 1998); Tony Clarke and Maude Barlow, *MAI Round Two: Global and Internal Threats to Canadian Sovereignty* (Toronto: Stoddart, 1998); for human rights issues consult, for instance, Amnesty International, Canadian Section, www.amnesty.org, or International Centre for Human Rights and Democratic Development, www.ichrdd.ca.

2. The Caledon Institute of Social Policy, "Speaking out project," *Occasional Papers*, Vol. 1 (Toronto, 1997); the Interfaith Social Reform Coalition, *Our Neighbours' Voices* (Toronto: Lorimer, 1998); Canadian Environmental Law Association, *Overview of Federal Law, Regulation and Policy* (Toronto, May 1998).

3. Sally Leaner, *Basic Income: A Primer* (Toronto: Between the Lines, 1999).

4. James Tobin, "Speculators' tax" in *New Economy* (Forth Worth: Dryden Press, 1994); Arthur Cordell, T. Ran Ide, Luc Soete, Karen Kamp, *The New Wealth of Nations: Taxing Cyberspace* (Toronto: Between the Lines, 1997).

5. For background on Citizens for Local Democracy visit C4LD's website: Http://community.web.net/citizens. The website contains archives as well as ongoing activities.

    For a practical guide to the mix of synchronous and asynchronous activities in community groups, see for instance, Maureen James and Liz Rykert, *Working Together Online* (Toronto: Web Networks, 1997).

6. Constance Mungall and Digby J. McLaren, eds., *Planet under Stress: The Challenge of Global Change* (Oxford: Oxford UP, 1990).

INDEX

The CBC Massey Lectures Series

The Politics of the Family
R. D. Laing
0-88784-546-0 (p)

The Educated Imagination
Northrop Frye
0-88784-598-3 (p)

The Real World of Democracy ˙
C. B. Macpherson
0-88784-530-4 (p)

Available in fine bookstores and at www.anansi.ca